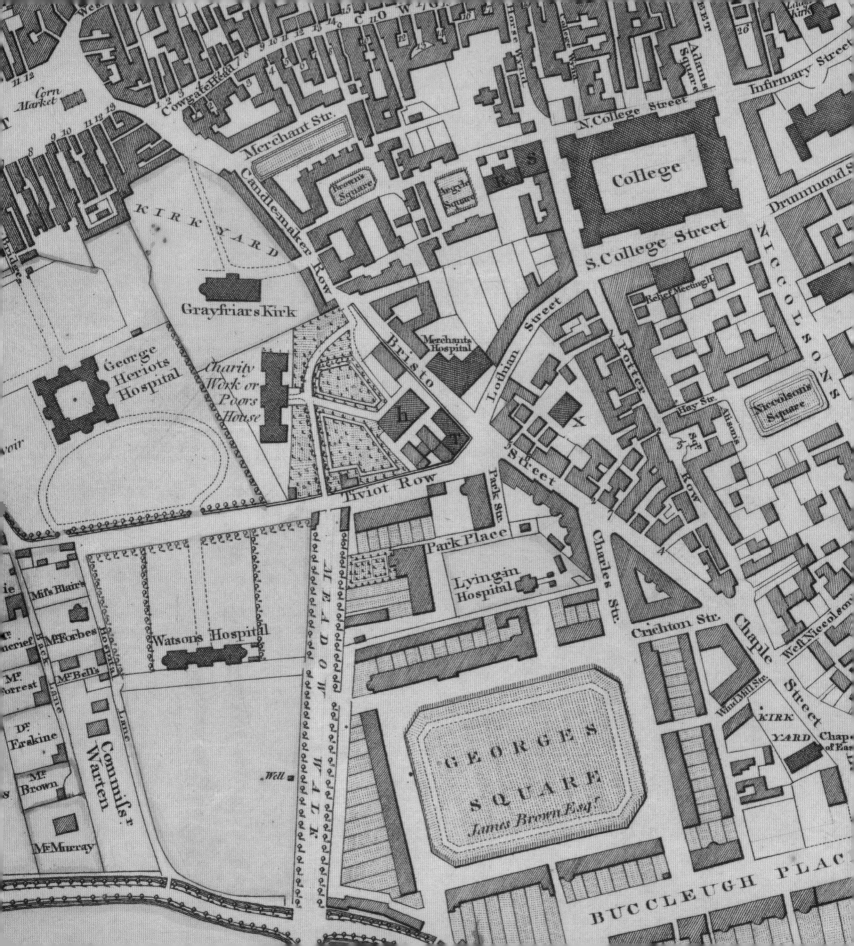

Edinburgh
MAPPING THE CITY

Christopher Fleet and Daniel MacCannell

BIRLINN

in association with the National Library of Scotland

First published in Great Britain in 2014 by
Birlinn Ltd

West Newington House
10 Newington Road
Edinburgh
EH9 1QS

www.birlinn.co.uk

3

ISBN: 978 1 78027 245 0

Copyright © Christopher Fleet and Daniel MacCannell 2014

The right of Christopher Fleet and Daniel MacCannell to be identified as the authors of this work has been asserted by them in accordance with the Copyright, Designs and Patents Act, 1988

All rights reserved. No part of this publication may be reproduced, stored, or transmitted in any form, or by any means, electronic, mechanical or photocopying, recording or otherwise, without the express written permission of the publisher.

British Library Cataloguing-in-Publication Data
A catalogue record for this book is available on request from the British Library

Typeset and designed by Mark Blackadder

Printed and bound by PNB, Latvia

Contents

	Introduction	XI
	Acknowledgements	XV
c.1530	The earliest printed view of Edinburgh	1
1544	A bird's-eye view of the Earl of Hertford's assault on Edinburgh	5
1560	A 'military platt' of the siege of Leith	9
1567	The murder of Darnley at Kirk o' Field	13
1577	The 'Lang Siege' of Edinburgh Castle, 1571–73	17
c.1582	Edinburgh on the European stage	21
c.1610	The earliest map of the Lothians	25
1647a	The most detailed bird's-eye view of the burgh	29
1647b	Viewing the Old Town from north and south	33
c.1682	A detailed manuscript map of Edinburgh's hinterland	37

c.1690	The 'Queen Anne' view from Calton Hill	41
1693	Greenvile Collins and the professionalisation of chartmaking	45
1710	'Le grand secret': re-fortifying Edinburgh Castle	49
1742	The first overhead plan of Edinburgh	53
1750	Post-Culloden Edinburgh Castle as a base for the Roy Military Survey	57
1752–55	The Roy Military Survey	61
1765	New public and private buildings	65
1766	Fact, fiction and Craig's original plan for New Edinburgh	69
1767	James Craig's proposals for the New Town	73
1773	The earliest buildings of the First New Town	77
1779	Land disputes around the Nor' Loch	81
1784	The suburbs on the South Side under construction	85
1790	'La Nature à Coup d'Oeil': Edinburgh's first panorama	89
1793	Old College and the new South Bridge	93
1804	Expanding to the north?	97
1813	The Canongate in transition	101
1815	The Forth in the Napoleonic era	105

CONTENTS

1817	'Containing all the recent and intended improvements …'	109
1819a	A curious double perspective on the east end of Princes Street	113
1819b	'There are no stars so lovely as Edinburgh street-lamps'	117
1822a	Disentangling past, present and future in Scotland's great county atlas	121
1822b	Leith at the time of King George IV's visit	125
1823	Announcing the Greek Revival	129
1826a	Parkland, estate mapping and suburbia	133
1826b	An early 'gated community': Newington	137
1832	Extending the franchise on paper	141
1834a	Maps for 'the masses'	145
1834b	A new harbour for the Age of Steam	149
1836	Development falters	153
1843	Holyrood Park: a new pleasure ground	157
1847	Foul burns and fertile meadows	161
1850	Neo-classicism soldiers on: plans and elevations for the Dean Estate	165
1851	The arrival of the railways	169
1852 / 1877	The Ordnance Survey's most detailed mapping ever	172

1852	The tourist city comes of age	177
1866	Clearing the Old Town slums	181
1869	The politics of drains and sewerage	185
1876	Park of wonders: the Royal Patent Gymnasium	189
1879	Mapping the speed of sound for the Time Gun	193
1886	A very Scottish international exhibition	197
1890	The Caledonian's tunnel along Princes Street	201
1891a	'The finest and most elaborate map of the city and suburbs ever produced'	205
1891b	Mapping a typhoid outbreak to track down its source	209
1892a	Tuberculosis as a window on late Victorian Edinburgh society	213
1892b	Mapping goes underground	217
1895	From city to conurbation: powering and extending the tramway network	221
1898a	Advancing into the light	225
1898b	The Usher Hall on the Meadows	229
1902	Civic survey, social evolution and the *Encyclopaedia Graphica*	233
1903a	Banishing markets and slaughterhouses	237
1903b	Triumph in the face of tragedy and governmental indifference: the Bathymetrical Survey	241

CONTENTS

1906	A Goad fire-insurance plan of the George IV Bridge environs	245
1919	A colour-coded chronology of Edinburgh's historical development	249
1923	Mapping an alcoholic 'bog of self-indulgence'	253
1932	The making of the Scottish Zoological Park	257
1934	Post Office Directory maps and urban history	261
1941	A German bombing map of Edinburgh	265
1946	RAF photography and Operation Revue	269
1949	Postwar visions for the motorcar age	273
1983	Mapping for a Soviet tank advance on Edinburgh	277
2014	Open, collaborative, volunteered digital data	281
	Epilogue	285
	Further reading	286
	Index	288

Introduction

All cities are unique and special, but some would seem to have a greater claim than others on the hearts and minds of visitors and their own inhabitants. Whether in terms of its topography and scenery, its architecture, its history, its townscape, its artistic and cultural life, its status as the capital city of Scotland, or some other attribute – perhaps indefinable – Edinburgh is one of the world's great cities. But it is also a city of contrasts, not just in terms of its famous Old and New Towns, but also as a place of danger, disease, destitution, violence and injustice amid splendour, beauty and achievement. Maps capture these themes and contrasts in a very direct and revealing manner.

In the past there has been a tendency to write cartographic histories as stories of progress or improvement, with a reassuringly triumphant subtext of better surveying, increasing detail, or greater accuracy as we march forwards through time. The reality is neither so neat nor so simple, but much more interesting. In the case of Edinburgh, viewed solely in terms of accuracy, the Ordnance Survey's heyday was very much the mid nineteenth century, whilst for those wanting to scrutinise the detail of all the wynds and closes in the Old Town, we move steadily downhill after James Gordon's 1647 bird's-eye view. For those interested in some other particular aspect of Edinburgh's history – its distributions of rich and poor, infrastructure, patterns of mortality, or underlying geology – the cartographic peaks and troughs would be different every time. Maps have never been an objective mirror of the real world, and their interest rests not just on their omissions and their 'cartographic silences' regarding features that are elided from our view, but also on what they proposed or imagined would happen, but never did. From these perspectives, maps that were *not* based on an original survey, or were *not* particularly noteworthy for their precision or detail, hold far more of interest to us, often revealing more than their creators intended about the social, political and economic forces that shaped Edinburgh's history.

This book gathers together a small selection of 71 maps that provide insights into the history of Edinburgh, reflecting the city at one level, and at the same time shaping our perception and understanding of it. We use a definition of 'mapping' that is deliberately broad, adopted from the current international History of Cartography project: 'graphic representations that facilitate a spatial understanding of things, concepts, conditions, processes, or events in the human world'. On this

---

H.M. Caddell / J.G. Bartholomew, 'The site of Edinburgh in ancient times',
Scottish Geographical Magazine, 9 (1893), between pp. 312 and 313

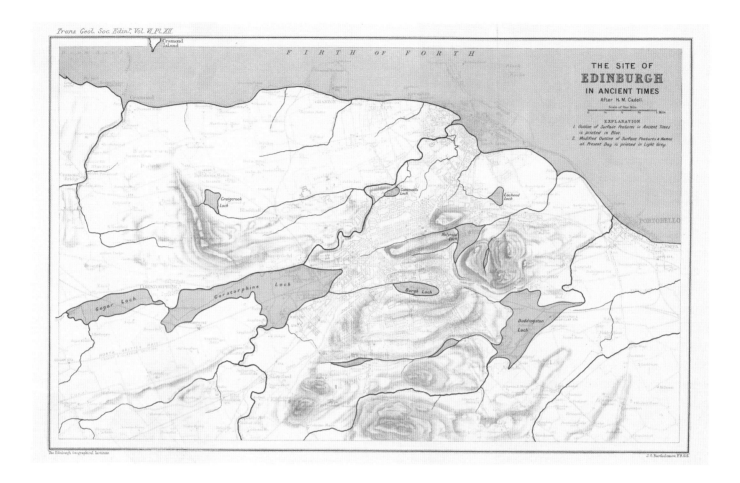

basis, we can happily include views, profiles, plans, sketches, and even an air photograph, focusing on their common threads and on what they *do* and *mean*, rather than their physical attributes or media. We also aim to situate these not only within a wider international history of mapping, but within specific local contexts. Close examination of English, German and Soviet military maps of the city informs, and is informed by, an awareness of geopolitics, warfare and strategy. Similarly, maps from the great European town atlas of Braun and Hogenberg of the late sixteenth century, or the magnificent world atlases of Hondius and Blaeu in the seventeenth, and even the more utilitarian parliamentary mapping associated with the Great Reform Act of 1832, can best be understood within their broader contexts. But maps also depict the uniqueness of Edinburgh's geography and history through specific events such as the arrival of George IV in 1822, the installation of the famous Time Gun in 1861, or the International Exhibition of 1886, as well as plans of individual buildings, streets and districts.

Maps can be read and understood at many levels, and part of their interest rests upon an admission that we may never completely understand them or fully capture their multiple meanings, especially as understood by people in earlier ages. Part of their attraction also rests upon their selective evidence: the making of a good map requires conscious and unconscious inclusion and exclusion of features from the real world. This process is not only culturally dependent, but continually changes over time; and so we often see Edinburgh altering more in the eyes of its beholders than on the ground. Between the Earl of Hertford's map of his 1544 invasion, James Gordon's flattering bird's-eye view of 1647, and the first overhead map of Edinburgh by William Edgar in 1742, the overall size and shape of the town alters relatively little, and yet these maps offer profoundly different presentations, reflecting their different makers, purposes and intended readership. Whereas today we are surrounded by maps in a myriad of forms, and the association between maps and travel appears perfectly obvious, it should be remembered that, until the Restoration period or even later, very few people had access to maps or would have perceived any need for such access, and the few who did were very unlikely to use them for route-finding.

Whilst graphical depictions of space are known to have existed from Roman times onwards, and a squashed island of Scotland appears on many medieval *mappae mundi*, Edinburgh is not named on surviving maps from the twelfth century or earlier. When it does eventually appear on maps, such as the Hereford *mappa mundi* or Matthew Paris's maps in the thirteenth century, the famous Gough map of the fourteenth century, or within John Hardyng's *Chronicle* of 1457, it is only as a name and symbol, giving no sense of its layout or appearance. This would change following the 're-discovery' of Ptolemy's *Geographia* and its publication in Renaissance Italy from 1477, which – allied with voyages of discovery, the spread of printing and literacy, and a greater awareness of the practical utility of mapping among statesmen and merchants – revolutionised the nature of global and large-scale mapping in Europe. Great humanist works such as Hartman Schedel's *Nuremberg Chronicle* from the 1490s and Sebastian Münster's *Cosmography* from the 1540s illustrated a large number of cities for the first time, and military engineers drafted detailed views and scaled drawings as a necessary part of attacking and defending them. Our earliest maps of Edinburgh date from this time, and historical maps like Henry Caddell's 'The site of Edinburgh in ancient times' (after the Ice Age), or Frank

F.C. Mears, 'Edinburgh, ca. 1460', *Scottish Geographical Magazine*, 35 (1919), p.313

Mears's 'Edinburgh, ca. 1460', whilst serving useful roles, are works of historical reconstruction and inevitably reflect something of their later compilation.

Within the roughly 485-year period for which any detailed maps are available, we have been keen to select maps that promote the special accomplishments of the people who made them, especially those who lived and worked in Edinburgh. Over the centuries, mapmaking employed a range of specialist skills – surveying, compiling, drafting, engraving, printing, and publishing – that were only rarely found in combination in particular institutions, let alone individuals. Georgian Edinburgh, however, quickly became a place where all of these skills could be found in relative abundance, earning it a special place in cartographic history. A number of leading surveyors – John Adair, John Laurie, John Ainslie and John Wood – lived in the city, which was also the centre of national cartographic initiatives including the Roy Military Survey (1747–55), John Thomson's remarkable *Atlas of Scotland* (1832), and Bartholomew's *Survey Atlas of Scotland* (1895 and 1912). From the eighteenth century onwards, Edinburgh also grew to become a major centre of expertise in engraving, and as this expanded to encompass lithographic printing in the nineteenth century, Edinburgh map publishers W. & A.K. Johnston and Bartholomew's became justly famous all over the world. The importance of Edinburgh as a centre of map production also reflected the city's unique academic, intellectual and political contexts, in which mathematicians, geographers, sociologists, engineers, geologists, physicians, public health officials and town planners all created and used maps (illustrated in this volume) as part of their work.

This book is necessarily selective, both in terms of the maps chosen, and in the details included of them, with the hope that it inspires further interest and exploration. The main carto-bibliography of printed maps of Edinburgh to 1929 records 195 maps, more than double the number included here, even before one starts to count the larger numbers of unpublished manuscript maps, maps within other publications, and map series produced by state-funded institutions. The selection process has of course been difficult, and ultimately personal, driven partly by the main themes, events and details of Edinburgh's history and cartographic history that each map illustrates, as well as the aesthetics of the maps themselves. As this book is published in collaboration with the National Library of Scotland (NLS), our bias too has been towards NLS's collections. We have aimed to provide balance and breadth through the inclusion of manuscript and unpublished mapping, and have sought to redress the chronological imbalance caused by the abundance of nineteenth- and twentieth-century maps compared with those from earlier centuries. Ordnance Survey maps are particularly numerous and valuable for the detail they show of Edinburgh from the 1850s through to the present day, but as they are relatively well known – and could easily have filled this volume to the exclusion of all else – they appear only a couple of times. Where possible, we have represented the broad, historically rural region that lies beneath the wider city today, even if the focus has necessarily been on the historic centre of Edinburgh and its port of Leith. We have also taken a decision to show only extracted details from most large maps so that they are legible and clear; complete zoomable versions of many of these maps can be explored online (see 'Further Reading', p. 285). In short, we are quite aware that it would have been possible to construct an entirely different book on the same topic; and can only hope that our pattern of inclusions and omissions does justice to this extraordinary city and its remarkable cartographic legacy.

Acknowledgements

We are particularly grateful to Hugh Andrew of Birlinn for his continuing support for publishing on Scottish historical cartography, and for encouraging the present work in particular. We are also indebted to the many excellent historians and researchers on Edinburgh and the Lothians and maps thereof whose work has been utilised within this volume (please refer to 'Further Reading'). We would also like to thank Carolyn Anderson, Ron Hill, Juliet Flower MacCannell, Hugh Salvesen, Adam Wilkinson and Rosemary Wake for reading and offering comments on earlier drafts, but we accept full responsibility for any errors that remain.

Unless stated otherwise, all images are reproduced courtesy of the National Library of Scotland.

Dates

The main date preceding each entry usually relates to the date of survey or compilation, to the extent that this is known. These main dates are also given in square brackets, e.g. [1866], within the text as cross-references to other entries in the book. The dates of publication and related bibliographic information are given in the caption relating to each image.

Septētrio

c.1530

The earliest printed view of Edinburgh

Edinburgh from the north, a woodcut after a drawing by exiled Scottish Lutheran theologian Alexander Allane (1500–65), has a good claim to be the earliest surviving printed view of Edinburgh. It first appeared in the 1550 Latin edition of Münster's *Cosmographia*, one of the most successful and popular books of the sixteenth century; the version shown here is from a French-language edition of 1575.

It is perhaps impossible to exaggerate the significance of the *Cosmographia* for European society: for the first time, a wide public could see the way the Earth looked through views and maps that were easy to read. Sebastian Münster (1488–1550) was famous in his lifetime as a theologian and teacher of Hebrew, but geography and cosmography were his true passions. Originally from Ingelheim near Mainz, he settled in Basel, where he converted to Protestantism in 1530 and gathered geographical information from a steadily growing circle of informants, including Allane. The *Cosmographia* was first published in 1544, and by 1550 (when Münster died from the plague) it had expanded to 1,233 pages, with 910 woodcut illustrations. Its text and images were both regularly updated under Münster's successors, and in addition to French, the work was translated into Czech, German and Italian.

Allane was born in Edinburgh and was one of the first students of St Leonard's College, St Andrews. He was forced to flee from Scotland in the early 1530s to escape persecution by Patrick Hepburn, Catholic archbishop of St Andrews, and spent several years in Wittenberg, the home of Martin Luther, as well as in Cambridge before becoming professor of theology at Leipzig in 1542. Following his dramatic escape from Scotland, he acquired the pseudonym Alesius. While Latin or (as here) Latinised Greek pseudonyms were commonplace in humanist circles, this one means 'exile' or 'refugee'. The eventual Protestantisation of Scotland proceeded along Calvinist, not Lutheran, lines, but Lutheranism – especially

Sebastian Münster, *Alexandre Alesie Escossois D'Edinbourg* (1575)

the form adopted in sixteenth-century Denmark and Sweden – remained a strong influence on the development of Scottish Episcopalianism in the seventeenth century and beyond.

Perhaps naturally, given his primary interests, Allane's view records a city very much dominated by churches and monasteries. Between Holyrood Palace (*A. Le palais du roy*) to the left and the Castle (*B. La tour des filles*) to the right, he singles out St Cuthbert's (*C. L'Eglise S. Cubert*), St Giles (*D. L'Eglise S.Gilles*), Greyfriars (*E. Les freres mineurs*), St Mary in the Field or Kirk o' Field (*F. L'Eglise S. Marie au champ*), Trinity College (*G. Le college de la royne*, so named after its foundation by Marie de Gueldres, wife of King James II, r. 1437–60), and the monasteries of Blackfriars or 'Friars Preachers' (*H. Les freres prescheurs*) and Holyrood (*K. La monastere S. Croix*). It is interesting to contrast Allane's view and its depiction of these buildings with Richard Lee's near-contemporary military sketch [1544], also from the north.

Topographically, we get a rough sense of Edinburgh's hilly site, and even Arthur's Seat and perhaps the Pentlands in the background, with the port of Leith and the Forth in the foreground, as well as a short section of crenellated wall near Holyrood. Otherwise, the features shown could depict almost any city in Europe. In part this reflects the necessary imprecision of Allane's original drawing, executed from memory after many years in exile; but the style, knowledge and enthusiasms of the Swiss woodcut engraver, Hans Deutsch (1525–1571),

c.1530

Kirk o' Field (F), Greyfriars (E) and the Castle (B).

Holyrood Palace.

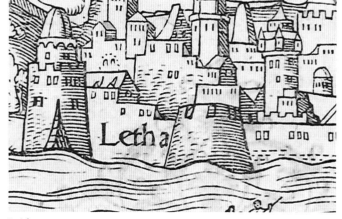

Leith.

also undoubtedly had roles to play. In particular, the buildings that are *not* churches – mostly towers with a variety of slanted, pyramidal and domed roofs – are more evocative of Renaissance northern Italy than of anything that existed in the British Isles in that, or any other, time period. Despite what we would see as its lack of correspondence with the real world, this work when new was 'state of the art', and the great majority of its readers would have had no first-hand knowledge of Scotland with which to compare it.

Within Edinburgh, the Reformation of 1560 brought profound, as well as some more subtle, changes to the religious buildings and institutions depicted here. Whilst the Greyfriars and Blackfriars monasteries were attacked, looted and left in ruins in 1559, the other ecclesiastical buildings depicted were altered for Protestant worship. St Giles became the main parish church, subdivided into three, while St Cuthbert's (the original parish church) and Trinity College Church survived for the northwest and northeast suburbs respectively. Kirk o' Field, following the murder of Henry, Lord Darnley [1567], survived to become the new town's college from 1583.

It should be noted that Allane's view is the only map in this book that was printed from a woodcut, explaining the bold line work, though the title and key names were inset in type. All the later printed maps of Edinburgh, up until the mid nineteenth century, were printed from copper plates.

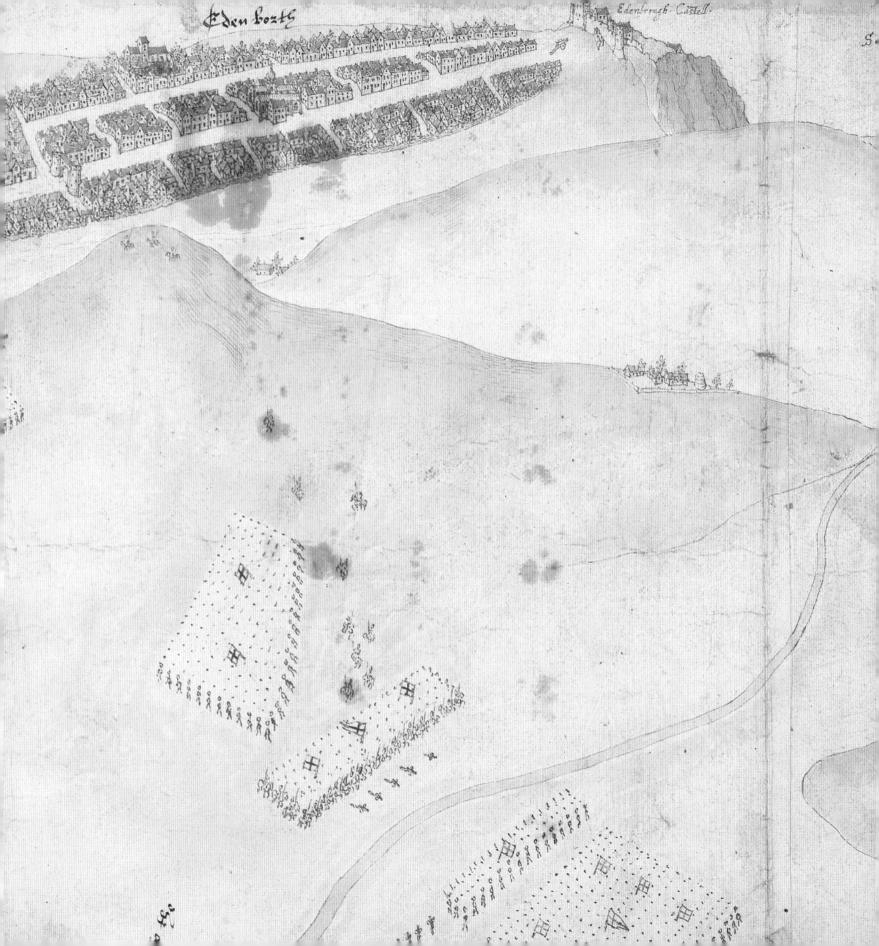

1544

A bird's-eye view of the Earl of Hertford's assault on Edinburgh

This famous bird's-eye view of Edinburgh from the northeast records the horrific assault on the town in April–May 1544 by the English army of Edward Seymour, Earl of Hertford. This was part of the 'Rough Wooing': a period of invasions, occupation and significant bloodshed that followed the Scots' rejection of a marriage that Henry VIII sought to impose between his son, the future King Edward VI, and the infant Mary, Queen of Scots.

The view is by Richard Lee (1501/2–1575), an English military engineer who accompanied Lord Hertford on the Scottish campaign. Lee's father and grandfather may have been masons, and by the age of thirty he had emerged as England's foremost expert on military engineering. He was appointed surveyor of Calais from 1536 and did much over the following eight years to maintain and expand the fortifications there, in one of the largest construction programmes in northwestern Europe since Roman times. It is now largely forgotten that Calais was central to both England's geopolitical strategy and its economy, with a mostly Protestant, English-speaking population of 12,000, making it the Tudors' third largest town, after London and Norwich. At Calais, Lee clearly impressed Hertford who personally selected him for the famous assault on the Scottish capital depicted here.

Surviving among Sir Robert Bruce Cotton's 'Augustus' manuscripts in the British Library, this image is a view of the English army entering Edinburgh, with troops marching under the English flag near Calton Hill in the foreground. The only text labels the Palace of Holyrood to the left ('the kyng of skotes palas'), Edinburgh itself ('Eden borth') and the castle ('Edenbvragh Castell') to the far right. Kirk o' Field appears as a large cruciform church with a central tower to the left of the Netherbow, whilst the formidable great hall and church

Richard Lee, Bird's-eye view of the town of Edinburgh (1544).
By permission of the British Library

of the Blackfriars appears to the right in the background; this may be the only surviving image of the Blackfriars' building before its destruction in the following decade. Further up the Royal Mile, St Giles (with the Tolbooth building in front) can also be seen, as can the Watergate Port at the foot of Leith Wynd. Other buildings and streets are largely sketched in conventionally, though interestingly the roofs of Edinburgh are picked out in red while those of the then-separate burgh of Canongate are not. This may have been an attempt to distinguish pantiled roofs in Edinburgh from thatched roofs in Canongate, but so few complete pre-seventeenth-century houses (or images of houses) survive in either burgh that assessing the general accuracy of this would probably be impossible. That being said, the dwellings shown here are considerably more believable, in point of both form and scale, than those depicted by Alexander Allane and Hans Deutsch [c.1530]. The surprisingly small extent of Edinburgh Castle in Lee's drawing, meanwhile, reminds us that almost everything visible on the castle site today is from after the 'Lang Siege' of the 1570s [1577], and indeed mostly from after the restoration of the monarchy in 1660, when Scotland's first standing army was established and the castle was expanded to serve as a garrison.

After landing at Leith, the English army marched on Edinburgh. It is worth noting the vast bulk of its infantry formations, which in no way anticipate the linear tactics of the ensuing centuries. Spain, flush with Inca gold, was still the dominant and most influential military power in the world, and something akin to the Spanish *tercios* – massive square blocks of up to 3,000 men – is indicated here. Edinburgh's provost, as mayors are still known in Scotland, offered to evacuate the town and give the English the keys, but this overture was quickly rejected. The attackers instead beat open the Netherbow Port, clearly shown here with its twin turrets and gate towards the lower left-hand slope of the Royal Mile, and proceeded with an attack on the castle (as symbolised by a single cannon). When the castle failed to yield, the invaders turned 'to ruynate and destroye the sayde towne with fyer' – ironically, burning buildings in so many places that the smoke forced them to quit the town and return for further pillaging

Holyrood: 'the kyng of skotes palas'.

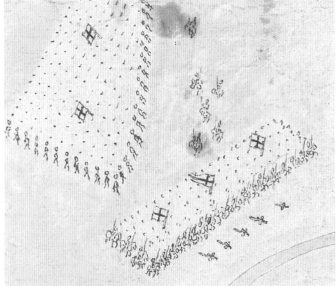

English infantry, upper left, and cavalry in formation carrying their St George's cross flags.

1544

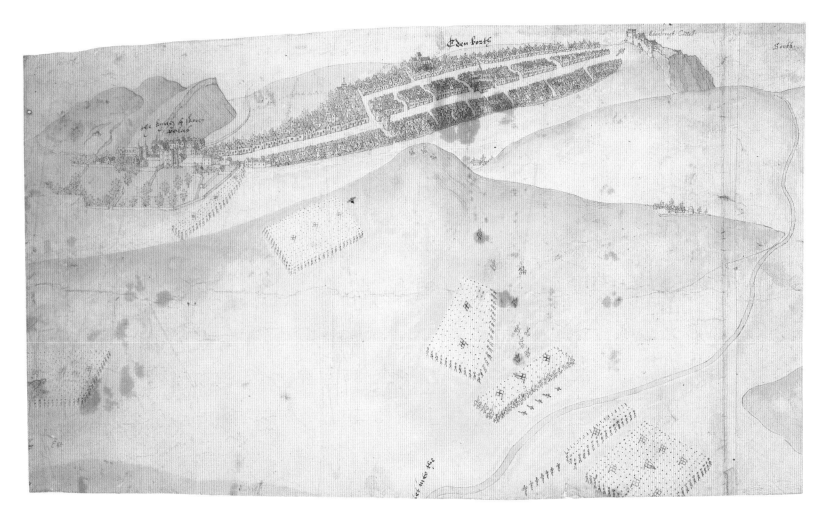

after the air cleared. 'Neyther within ye waules, nor in the suburbes, waas lefte any one house vnbrent'. Holyrood Abbey was also invaded, and Hertford's fleet attacked other towns and villages along the Forth from Fife Ness to Stirling.

Lee joined in the general pillaging, making off with two notable artefacts from Holyrood Abbey: an eagle lectern (said to have been a gift from Pope Alexander VI, r. 1492–1503) and a huge bronze font. Untypically for an early modern looter, however, Lee gave the lectern to his local church, St Stephen's in St Albans, and the font to the abbey church there. He was rewarded with a knighthood on 11 May 1544 for his part in the Edinburgh expedition and elected to Parliament the following year, but this did nothing to impede his illustrious and highly lucrative career as a military engineer. He designed the state-of-the-art *trace italienne* anti-artillery fortress at Eyemouth in 1547 and later the fortifications at Berwick-upon-Tweed, which constituted the grandest Renaissance fortress ever built in the British Isles and possibly the costliest single project of the Elizabethan age. Sir Richard could never have guessed, whilst managing a border-defence project worth £14 billion in today's money, that in less than half a century the border between Scotland and England would cease to be defended altogether.

1560

A 'military platt' of the siege of Leith

Tactically inconclusive but strategically pivotal, the siege of Leith brought to an end the French military occupation of Scotland that had begun in 1548. Some have also argued that it ended an attempt by English forces to establish a defensive zone or 'pale' within Scotland, comparable to the Pale of Calais or its more famous variant around Dublin. This wide-angled perspective view from the archives of Petworth House in Sussex records the detailed positions of mines and artillery on the day of the French defenders' capitulation, 7 July 1560. Petworth was owned by Sir Henry Percy, 8th Earl of Northumberland, who distinguished himself at the siege.

Following the bloody and decisive English victory at Pinkie Cleugh near Musselburgh in 1547, the Scots looked to France for assistance. From 1548, French forces arrived, eventually numbering 8,000 men and concentrated in Haddington, Broughty Castle and Leith. The Italian military engineer Migliorino Ubaldini, who also designed the Spur at Edinburgh Castle, planned a bulwark at the Leith Kirkgate around this time, but it was left to his more famous compatriot Piero di Strozzi (d. 1558) to continue his work. Strozzi had been shot in the leg at Haddington, and apparently supervised the construction of Leith's fortifications from a chair carried by four men whilst recovering. Leith became one of the earliest examples in Britain of the *trace italienne* style of fortification, with low, thick walls designed to absorb rather than resist cannon-fire, and the large, angled bastions illustrated here, which gave covering fire to one another. A number of historians contend that the *trace italienne*, owing to its unprecedented construction cost as much as to the strategic advantage it gave to the defending side, was the single most influential technology of the sixteenth and seventeenth centuries. Certainly, in this case, it proved a tough nut to crack.

In April 1560, English troops began constructing trenches and siege works around Leith, assisted by Scots Protestant

---

Richard Lee?, *The Plat of Lythe w' th'aproche of the Trenches Therevnto* (1560). Courtesy of West Sussex Record Office

forces levied by the Lords of the Congregation, a group of aristocratic revolutionaries who favoured John Knox's form of religion and co-operation with (Protestant) England rather than with Scotland's traditional (and Catholic) ally, France. The siege works begun in the east near Lochend and Restalrig, shown at centre left, were extended westwards to Pilrig by late April, and reached Bonnington by the Water of Leith (centre right) in early May, a length of two miles. The English also constructed two temporary 'fortlets', each about three acres in extent and both square with four corner bastions. These were known as Mount Pelham, to the left/east of Leith, and Mount Somerset, just above/south of the beleaguered town. The Mounts were built 13 feet above the surrounding ground, which helped their artillery to fire into Leith at ranges in excess of a quarter mile, near the maximum range of a 5.5-inch culverin.

The siege came to a head early on the morning of 7 May, and although there were two breaches in the walls, the scaling ladders proved too short and the English were defeated. In all, several hundred and perhaps as many as 1,500 Scots and English were killed, and many more injured. Nevertheless, the noose tightened over the following month, with mines being dug towards the fortifications; packed with explosives, these tunnels would have been detonated to create further gaps in the French defences if another assault had been made. In the event, however, peace talks culminated in the Treaty of Edinburgh, which secured the withdrawal of French *and* English troops from Scotland, and by 17 July the foreign soldiers had departed: a fitting epitaph, perhaps, for a fortress-town that could neither be successfully stormed nor indefinitely held. (Horse and rat meat were reportedly among the treats on offer near the end.) Captain William Pelham, the engineer for whom Mount Pelham was named, was present at the siege of Leith and would go on to help build the prodigious defences of Berwick-upon-Tweed. He later defended the English Pale in Ireland against the Fitzmaurice rebellion, for which he was knighted in 1579.

This map has been ranked as one of the most immediate of all records of the siege, and is an interesting combination of a scaled plan in the foreground with a pictorial bird's-eye view in the background. In the mid sixteenth century, the new style of overhead scaled plan was being introduced into Britain by the Italians, but took time to gain acceptance, and the conventional bird's-eye view long remained the more familiar and accessible mode of presentation. A note on this plan states that 'The Scale of this Plat [as maps were usually then known] is eighty paces to ane ynch. Every pace conteyning 5 foots geometricall', confirming its scale of 400 feet to the inch, or 1:4,800. Although we cannot be certain of its author, we know that Sir Richard Lee, England's chief military engineer, was present during the siege and he sent 'a platt of Leith' to London on 15 May, requesting the young Queen Elizabeth's decision regarding 'works' shown on it. Lee had been in Edinburgh sixteen years earlier [1544], and had also returned to Scotland in the interim to plan the *trace italienne* fortress at Eyemouth. Military engineers like Lee would have been required to draft maps such as this one as a key means of directing siege operations, perhaps assisted by local informants for specific details, as well as by subordinates like Pelham and Captain Francis Somerset.

Apart from the excellent detail shown of Leith's defences and the Anglo-Scottish siege works, the map confirms the existence and location of several significant buildings, including Restalrig Church (demolished 1560) and Restalrig Old Tower (destroyed 1586) on the far left. In the mid-right distance, by a bend in the Water of Leith, is the enclosed garden of Inverleith, whilst to the right Newhaven and Granton are both shown and named. The principal buildings in Edinburgh – Holyrood, St Giles and the Castle – are shown in a deliberately enlarged and stylised manner, with 'Cragge Ingalt' obscuring the Canongate; this particular presentation of Arthur's Seat and Edinburgh has some similarities to the surviving views of the Battle of Pinkie Cleugh, again attributed to military engineers. Although the Treaty of Edinburgh required the dismantling of Leith's fortifications, they are shown quite clearly, looking substantially intact, in the background of Rowland Johnson's view [1577].

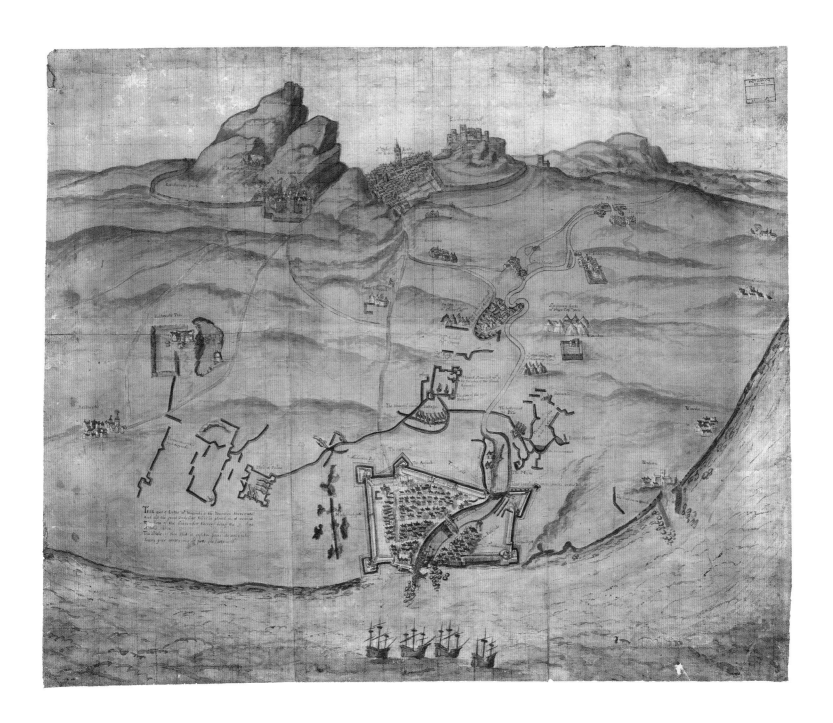

1567

The murder of Darnley at Kirk o' Field

The brutal killing of Henry Stewart, Lord Darnley, who was second husband to Mary, Queen of Scots, remains one of the greatest unsolved Scottish murder mysteries. This unique map, held in The National Archives of the United Kingdom, was drawn shortly after the murder for William Cecil, 1st Baron Burghley. Burghley was the chief advisor to Elizabeth I for most of her reign, and took a very keen interest in maps. English interest in the murder was likewise intense, both inside and outside governmental circles, especially since the victim was Elizabeth's cousin and former suitor. It is the measure of his unpopularity that, even today, he is universally referred to as 'Darnley' rather than King Henry (his official style from July 1565 until his death), and that he is always said to have been murdered rather than assassinated.

Henry and Mary's relationship was fraught with difficulties. He was handsome and charming, but also arrogant, vain and deeply jealous of Mary's other affections, and this culminated in the murder of David Rizzio, Mary's private secretary, in March 1566. Darnley was clearly implicated, and over the following months Mary colluded with her Lords of Council in actively trying to remove him.

On the night of 9–10 February 1567, Darnley was unwell, having contracted either smallpox or (as seems more likely) syphilis, and was staying in a house belonging to the brother of Sir James Balfour in the quadrangular former abbey of Kirk o' Field, which was located on the site now occupied by Edinburgh University's Old College. Mary visited her husband in the evening and then left to attend a wedding celebration, but in the early hours of the morning, the house was completely destroyed by a massive explosion. Curiously, Darnley's semi-naked body and that of his servant William Taylor were found under a tree in the orchard, accompanied

Anon., Bird's-eye view of the murder of Lord Darnley at Kirk o' Field church and churchyard (1567). Courtesy of The National Archives by a chair, dagger, coat and cloak, but with no visible marks of the explosion on them. But neither was there any evidence of death through strangulation, stabbing or blows to the head. On the whole, it seems most likely that they were suffocated. Shortly before Darnley died, neighbours claimed that he had cried out 'O my brothers, have pity on me for the love of him who had mercy on all the world', which has been interpreted as a possible fruitless appeal to his Douglas kinsmen who may have conspired against him. Archibald Douglas's shoes were found at the scene, and he was one of several supporters of James Hepburn, 4th Earl of Bothwell, around whom suspicion for the murder quickly began to revolve. Certainly, Bothwell was intimate with Mary, and he married her in May 1567, having divorced his first wife 12 days earlier. Although Darnley was unlamented, the murder shocked people of every political stripe, and led ultimately to Mary's abdication in July and her flight to England the following year.

This curious, semi-allegorical drawing is quite difficult to

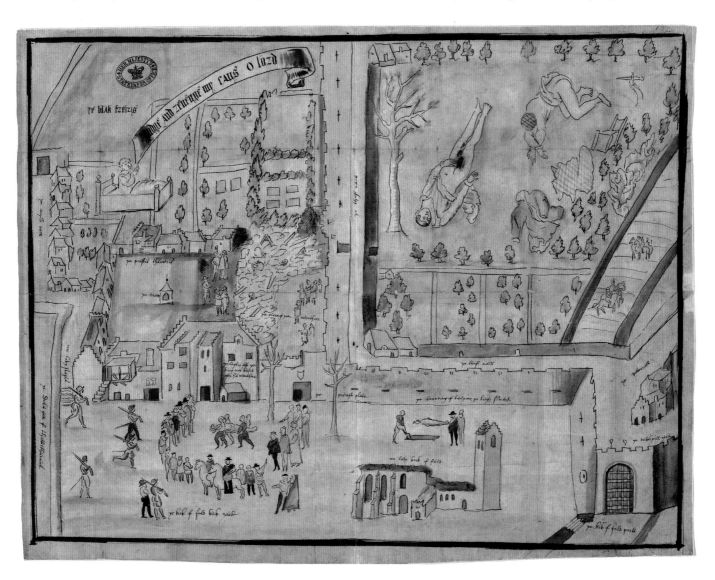

Darnley's body being carried away through the Kirk o' Field churchyard.

The burying of Darnley's servant behind the church.

interpret, for instance showing the Flodden Wall turning at right angles near the centre when in fact it was a straight line running along Thieves' Row. Kirk o' Field lay beyond the King's Wall (the town wall of the fifteenth century), but within the recently completed Flodden Wall. The large crow-stepped building at lower left is the Provost's House 'where the King was keepit after his murther'. Just above this, in the centre of the drawing, is 'ye place of ye murther' with roof-trusses and other debris strewn about, indicating the ruins of the Prebendaries' Lodging, the house that had been blown up. The adjoining door led into 'Ye Thieves Raw', and just opposite, on the south side of the road, is the doorway to the walled orchard or garden in which the bodies of Darnley and his servant were found. On the far left is 'Ye Mylk Raw' leading from Blackfriars Wynd and the Cowgate to 'Our Lady's Stepis' and on into the churchyard of Kirk o' Field. 'Our Lady Kirk of Field' itself is shown to the lower right with 'ye burying of body of ye King's servant'; while further to the right is Kirk o' Field port leading to Potterrow, with the Flodden Wall (correctly this time) turning to the south at this point.

Henry and Mary's infant son, the future James VI of Scotland and I of England, is shown in bed at upper left with a bold 'voice bubble' reading 'Judge and revenge my caus O Lord' – a sentiment echoed word-for-word on the banners of the rebel Confederate Lords who captured Mary at the Battle of Carberry Hill near Musselburgh on 15 June 1567 and forced her to abdicate, installing one-year-old Prince James as king in her place. The ensuing civil war between the 'Queen's Men' and the 'King's Men' [1577] would last more than five years.

Taken together with the arguably disrespectful depiction of Darnley's body, the presence of the Confederate Lords' motto suggests that this map, while produced for a foreign power, was in effect a platform for the views of Scotland's own anti-Marian faction. Be that as it may, Henry Darnley as a descendant of Henry VII was an heir to the throne of England, as well as being a Roman Catholic; so the childless, Protestant ruler of England would have gained a measure of relief from his demise, whether she was ultimately responsible for it or not.

Though no-one today strongly endorses their guilt, five associates of Bothwell were brutally executed for Darnley's murder. Bothwell was acquitted of the crime in Scotland after a trial lasting seven hours, but was still wanted for it in England. After Mary's abdication he fled to the kingdom of Denmark–Norway, only to be imprisoned in chains, possibly on English instructions. He died insane ten years later, aged 44.

1577

The 'Lang Siege' of Edinburgh Castle, 1571–73

This action-packed view of Edinburgh Castle under assault first appeared in Raphael Holinshed's *Chronicles of England, Scotland, and Ireland* in 1577. This was the leading Elizabethan history book – a monumental two-volume work, comprising 2,835 folio pages – which later became a major source and inspiration for the plays of both William Shakespeare and Christopher Marlowe. This particular Holinshed view was almost certainly based on an original 'platte' of the city drawn four years earlier by the military engineer Rowland Johnson, Surveyor of Works at Berwick-upon-Tweed.

The 'Lang [long] Siege', lasting nearly two years, made Edinburgh the main focal point of the bitter civil war that had been fought since 1568 between the Queen's Men – who supported the deposed Catholic Mary, Queen of Scots – and the King's Men, supporters of her Protestant son, James VI. The siege was triggered by the murder of the regent James Stewart, 1st Earl of Moray, a moderate Protestant and Queen Mary's illegitimate half-brother. With Moray out of the picture, the Protestant faction's tendencies toward Anglophilia and religious extremism both accelerated. Probably as a result, Moray's close friend Sir William Kirkcaldy of Grange – who had been the Governor of Edinburgh Castle for several years – changed sides, joining the Queen's Men and taking the castle and its garrison with him, along with the crown jewels, the state papers and the nation's single largest collection of heavy artillery.

Led by the ginger-bearded 4th Earl of Morton, who served as the boy-king's regent, the now-outgunned King's Men were headquartered in Leith, as indicated here by a group of 12 military tents in a bend in the Water of Leith, apparently in the modern vicinity of Great Junction Street. The two-year local stalemate between the two Scottish parties was only broken by the arrival, in April 1573, of an English force of heavy artillery and soldiers under Sir William Drury, following an

---

Rowland Johnson?, Siege of Edinburgh Castle (1577)

appeal for help that was sent to Elizabeth I by the King's Men.

The view gives a deliberately flattering portrayal of the English besiegers, identifiable by their flags with the cross of St George. Also depicted are their Scottish allies, the King's Men under their saltire banner, who fire eastward into the castle from the 'King's Mont', a temporary fortification in the present-day vicinity of the Usher Hall. Both sides barricaded the High Street at various times, the Queen's Men clearing the western end of dozens of houses in an attempt to obtain firewood as much as to create a clearer field of fire. In early 1573, a total of three ramparts or traverses had been formed in the High Street, including a substantial one in the Overbow, the old west port at the head of the Lawnmarket, and these helped to reduce the number of casualties amongst the townspeople (who numbered 10,000 within the walls at this date). From 22 May, when the English artillery was deployed, around 3,000 shots were fired at the castle. David's Tower and the Constable's Tower, both built in the fourteenth century, were quickly reduced to rubble, blocking the castle's main entrance and one of its wells. In this view, David's Tower has not yet been destroyed, and can be identified from its three prominent Gothic windows, as well as its greater size relative to the other towers.

The view also clearly picks out the famous 'Spurr' – a state-of-the art projecting defensive work designed by Ubaldini [1560] and constructed in 1548. The Spur was 20 feet high, protected 'with turfe and basketes, set up and furnished with ordnance', and significantly delayed the surrender of the castle. At 7 a.m. on 26 May, however, the Spur was successfully assaulted as illustrated here (the non-destruction of David's Tower notwithstanding), and the following day Grange negotiated for a surrender. When it became clear that he would not be spared, Grange tried to continue the resistance, but the beleaguered garrison threatened to mutiny. When Drury's men entered the castle on 28 May, most of the garrison were in fact allowed to go free, but not Grange, who was imprisoned for more than two months before being hauled through the streets of Edinburgh behind a cart and executed at the Mercat Cross – thus fulfilling a famous (and perhaps self-fulfilling) prophecy made by his great enemy, John Knox.

Grange, 'a lusty, stark and well-proportioned personnage, hardy and of a magnanimous courage', was a Protestant and a rather unlikely champion of the Marian cause. Having participated in the assassination of Cardinal David Beaton in 1546, he worked as a secret agent for Protestant King Edward VI of England in the early 1550s, and personally took the surrender of Queen Mary after the Battle of Carberry Hill in 1567; he then defeated her again at Langside (in the face of superior numbers) the following year. Grange appears to have been motivated throughout the civil war by a then-unusual combination of firm Protestantism and a suspicious dislike of foreign intervention in Scottish affairs, particularly by the English.

Quite apart from the fact that it definitively ended Scotland's five-year war of religion, the siege of Edinburgh Castle was a major news event, becoming the subject of a broadside ballad, *The Sege of the castel of Edinburgh* (Edinburgh, 1573). Printed shortly after the siege – and so-called 'news' ballads often took several years to appear – this contained unique, eyewitness information and may be considered a major milestone in the development of news reporting in the British Isles. Its publisher, Robert Lekpreuik, had also produced eight different ballad accounts of Moray's assassination, all within the year it occurred; and he appears to have been among the first people in Britain to print news of France's St Bartholomew's Day Massacre, which took place in the summer of 1572, more or less in the middle of the Edinburgh siege.

A detailed contemporary account of the siege was also written by Thomas Churchyard, who accompanied Drury on the expedition to Scotland. Churchyard had fought at Pinkie Cleugh in 1547, laid waste to the East Neuk of Fife the following year, and was present at the 1560 Siege of Leith. His poetry, like the map seen here, embellishes and glorifies what must have been a violent and messy affair on the ground:

This lofty seat, and lantern of that land
Like lode starr stode, and lokte o'er ev'ry streete
Wherein there was a stout sufficient band

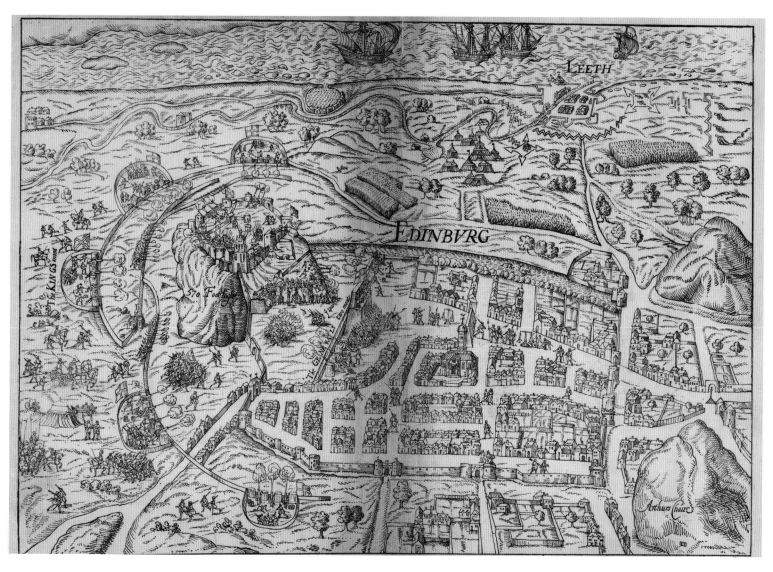

That furnisht were, with corage wyt and spreet
And wanted nought, that serd for their defence
Or could in fine, repulse their enmies thence
Well storde with shot, yea sure both good and great
That might far of, at wyl the countrey beat.

The houses shown here are more detailed and more various in form than those depicted on Lee's map of 1544, but are still essentially stereotyped, with occasional small wallhead dormers being perhaps the main concession to reality. Apart from the destruction of an entire neighbourhood in the vicinity of the Spur, the most direct architectural response to the siege was the 1574 construction of the Half Moon Battery on the former site of David's Tower, as clearly shown on later maps [for example, 1750]. The fact that only the Flodden Wall is depicted seems to reinforce comments in the written record, to the effect that the fifteenth-century King's Wall had long since been obscured by housing built on and around it.

c.1582

Edinburgh on the European stage

This 1657 Amsterdam edition of Georg Braun and Franz Hogenberg's famous bird's-eye view of Edinburgh from the south is a light re-working by Johannes Janssonius (1588–1664) of an earlier view, engraved around 1582 for inclusion in the *Civitates Orbis Terrarum*. The *Civitates* was published in Cologne between 1572 and 1617 in six volumes, and included 546 prospects and views of European towns. This great city atlas, edited by Braun and largely engraved by Hogenberg, was greatly influenced by Abraham Ortelius's *Theatrum Orbis Terrarum*, arguably the world's first true atlas, and it attempted to provide a similarly comprehensive collection of maps with a similar format and style. The views were copied and altered by later publishers, and this particular version omits the four large human figures that decorated the foreground of the original, as well as replacing the abstract mannerist strapwork design of the legend at lower left with an array of vaguely neo-Roman military trophies.

The inclusion of Edinburgh within the *Civitates* reflects something of its importance amongst European cities of the sixteenth century; and although a fine bird's-eye view of St Andrews was drafted in the 1580s, perhaps for the *Civitates*, Edinburgh was the only Scottish town included in the published version. It should be recalled here that, in the Middle Ages, Scottish kings toured the country constantly: hunting, collecting 'in-kind' taxation, visiting holy sites and dispensing justice personally. Parliament likewise sat in a surprising array of places including Aberdeen, Dunfermline, Inverness and Linlithgow, and had no special building in Edinburgh prior to the 1630s [1647a]. Given these cultural circumstances, a single capital city – or even the notion that the 'seat of government' would *be* a city – was somewhat alien. Edinburgh won the title of capital slowly, belatedly and haphazardly. Nor was

---

Johannes Janssonius, *Edenburgum vulgo Edenburg*, from
Illustriorum principumque urbium septentrionalium Europae tabulae (1657)

it always the pre-eminent town in medieval Scotland, being assured of this distinction only when its larger and wealthier rival, Berwick-upon-Tweed, was lost to England.

In many ways, this map tells us more about the publication and audience for whom it was intended than about Edinburgh. Its aesthetics and content reflect the need to create a pleasing image for the affluent European market, with neat ordered streets surrounded by colourful fields, and with limited specific real details of Edinburgh. It is very obviously based on the Holinshed view of the 1573 Siege of Edinburgh [1577], even to the extent that David's Tower – destroyed in the siege – is still present, while the Half Moon Battery that replaced it in 1574 is not. Likewise, the absence of Holyrood Palace from both maps may reflect the fact that the siege did not extend that far east, though at least in the earlier map the palace was off the right-hand edge of the page and not just magically gone, as here.

All that being said, a number of specific features are clearly shown: the castle and its fortifications are particularly impressive, even if outdated and embellished. The origins of the name *Castrum Puellarum* or Maidens' Castle are unclear, but according to one legend it can be traced back to a belief that virgins had been held captive there by the Picts, a group of Celts who ruled the area until supplanted by the Anglo-Saxons in the seventh century AD. Town walls are particularly prominent, and although stylised, the position and size as well as number of the six ports into the town are accurate enough. A prominent and clear Nor' Loch guards the northern side of

the town, and the larger churches, St Giles and Kirk o' Field, are shown roughly in their proper places. Although Holyrood is missing, the Girth Cross is clearly shown, significantly some way in front of the Abbey. By the time of the next detailed map of Canongate [1647a], additional buildings and streets would extend significantly further east, beyond the Cross, perhaps reflecting the flowering of the royal court and its growing need for accommodation during James VI's residence until 1603.

The accompanying text relating to Edinburgh deliberately flatters the town, whilst also bearing some grains of truth:

> To the west of the city a mountain with a tall cliff rises up, upon which stands a castle, which is extremely secure due to its natural setting. On the side of the city to the east lies a majestic monastery to the Holy Cross, alongside it a royal palace set within an exceedingly charming garden. There are two large streets in this city, one that is paved with ashlar stone, leading from the Maidens' Castle to the monastery and to the royal palace, and the other is the highway. From the highway running from north to south countless smaller roads branch off, all of which are lined with very tall buildings.

There was no space for squalor, disease, destitution, crime or even religious strife in the narrative.

The *Castrum Puellarum* or Maidens' Castle. Note the absence of the Half Moon Battery, constructed more than 80 years before this map was published, reflecting this view's origins in the Lang Siege [1577].

The eastern end of the Nor' Loch, showing the Flodden Wall and the Watergate and Netherbow Ports.

Kramond ynch

Lyth Roade

cle B.
Mainshou
Kottmoore
East Kragy
ragy hal
Bridgs end
Nether Kramont
Grantoun
Weirdy
Laurencetoun
Newhaven
Nether Bartoun
Pilmuse
Moore hous
Bonitou
Over Kramons
W. Dryley
Warestu
Lyth
over Barntoun
E Dryley
Canon mill
Pilrig
Restalrig
Nether Lany
Kammock
Inner Lyth
Broughtoun
Over Lany
Gracruik
Silver mill
Southfield
Revelstoun
Kratg
E. Craigs
Den mills
Cots
N. eedow sied
Coldham
Hill
Haus
Wright hous
West craigs
New bridge
Holirudhous
Korstorphin
Darry
Edenburgh
Nagas
Easter
Korstor. C.
Sauchtoun
Marchistoun
Arthur seat
Brunst
Nether Gogar
Scheens
The Parck
Wester
Gogar mil
Sauchtoun
Gorgy
Tipperlin
Craighous
Duddistoun
Brome house
Grange
Blackfoord
Hermistoun
Luchna Burn
Gorgy mill
Priestfield
Rydhewes
Kainron
E. Hales
Kray
Rydd hall
Nyddry
Balbertoun
Ouer Brail
Marshal
Ricartoun
Wester Hales
Oxtoun
Bridgerd
Craigmiller
Keyrhill
Burn house
Cohntoun
Nether Libertou
Woodhall
Over Libertou
Foulbr
Coldheun
Mortoun hall
North hous
Edmondsto
Bonely
Osty wel
Wauestoun
Mortoun
Store hous
Newtou
Curry
Over Curry
Southhous
Killyth
Stratoun
Wowmet
Curry hill
Hilend
Stratoun
Stratoun hall
Dalmahoy
Lumphoy
Long Gilmoortoun
Pentland hill
Pilmoore
Lyips
Penthland
Guters
Byrney
Cotslak
Grange
Pala
Harelaw
Roslyin more
Dalketh
Logannes

PYRE

OF

c.1610

The earliest map of the Lothians

Some time between 1583 and 1614, but probably before he became the parish minister of Dunnet in Caithness in 1601, a St Andrews graduate named Timothy Pont made a monumental survey of nearly the whole of Scotland. His detailed maps of the country's regions are the earliest available, and include a wide range of man-made and physical features: bridges, mines, tower-houses, castles, kirks and mills, as well as the expected towns and villages, hills and mountains, woodland, rivers and lochs. Though often controversially described as the first thing of its kind undertaken in Europe, Pont's work found contemporary parallels in many other countries, as nations and national boundaries achieved greater definition, and descriptions of places and their qualities – whether in words or images – were assigned great value by monarchs and statesmen as well as by the emerging merchant and professional classes. Whereas maps of Scotland before Pont usually mentioned fewer than 200 places, Pont's work records more than 20,000 place-names.

In spite of, or perhaps because of, this staggering level of detail, Pont had great difficulty finding partners who would print and publish his maps. Scotland's tiny printing community lacked the capital and technology for any such undertaking, which therefore had to be sent abroad: in this case to the Low Countries, which by the seventeenth century had emerged as the world centre of map publishing. This map of the Lothians is quite special in being the only work by Pont that was engraved during his lifetime; this was done around 1610 by the Amsterdam-based publisher Jodocus Hondius. We know too from the inscription below the title cartouche that the engraving process had been sponsored by Andro Hart, a prominent Edinburgh bookseller, bookbinder and printer who had premises on the north side of the Canongate near the

---

Timothy Pont/Jodocus Hondius, *A New Description of the Shyres Lothian and Linlitquo* from
Gerardi Mercatoris – Atlas sive Cosmographicae Meditationes de Fabrica Mundi et Fabricati Figura (1630)

RIGHT. The dedication to King James VI and I, as the 'Serene, Powerful James I, King of Great Britain, France and Ireland', his preferred style. The medieval English claim to France was finally dropped in 1800, and with it the fleurs-de-lys in the British arms.

OPPOSITE. Detail of the burgh gibbet, between the enclosed grounds of Grange to the west and Holyrood Park to the east.

Market Cross from 1610 to 1623. Despite Hart's local clout, however, there was an approximately 20-year delay between the engraving of the map and its publication, in 1630; and it was Hondius's rival publishing house, Blaeu, that eventually managed to produce all of Pont's maps – but not until 1654, four decades after the cartographer had died.

Although Hondius's engravers would have selected and generalised from Pont's original hand-drawn map of the area (which does not itself survive), they did this rather more carefully than Blaeu's. The detail of Edinburgh clearly picks out the castle on its hill; prominent buildings such as St Giles and Holyrood; the walls around the town, palace and abbey; and the approximate size and shape of the burgh. More significantly, in comparison to all preceding maps in the present volume, the surrounding rural hinterland is also packed with detail. Here are tower-houses, castles and larger residences whose names we can recognise today in the suburbs and villages that they once owned as pasture, field and forest. The trend amongst the nobility of enclosing areas of parkland around their estates – though further encouraged by legislation of 1616 – was already well underway, as we see at Corstorphine, the Grange and Craigmillar, as well the royal hunting ground of Holyrood Park, enclosed by James V (r. 1513–42). The principal rivers – the Almond, Esk and Water of Leith – appear prominently along with their bridges, as well as several lochs that would be drained in the following centuries. The map also praises the agricultural wealth of the region: 'this province is self-sufficient in everything needed for the use of the inhabitants: for it abounds in crops of all kinds, such as wheat, winter-wheat, barley, oats, etc.'

One of the many interesting details that can be spotted is the Burgh Muir gibbet, just to the right of the enclosed grounds

of Grange House. From 1586, when the Town Council started feuing parts of the Burgh Muir, the gibbet was moved further along Dalkeith Road to near the end of what is now East Preston Street. Hondius records faithfully Pont's original sketch showing two stone pillars supporting a crossbeam from which criminals and other miscreants were hanged. They were joined by the chained and sometimes dismembered corpses from other execution sites in Edinburgh, and a stone wall had to be constructed to keep out scavenging dogs.

The accompanying text to Blaeu's version of the Lothian map records 43 mills along the Water of Leith, several of which are shown here. Seven major routeways are shown radiating from Edinburgh, all of which form the major arterial routes today. Prior to the publication of John Ogilby's *Britannia* in 1675, roads were hardly ever shown on maps, which were not yet intended as aids for travel; so their inclusion here is unusual and suggests that they may have had a special significance in the Edinburgh area. Only a very few other Hondius or Blaeu maps show roads in other parts of Scotland, though it is a matter of some debate whether this implied an actual absence of roads, or merely a lack of interest in them on the part of Pont. Certainly, the state of the Scottish roads leapt to a place of central importance in government thinking during the very belated visits to their northern kingdom undertaken by James VI and I (in 1617) and Charles I (in 1633).

As with Braun and Hogenberg's Town Atlas [c.1582], this map would have been seen by a small minority of European society – royalty, nobility and the merchant classes – as a means of visualising the Lothians from afar. It presents a stable, ordered, selective landscape consisting of features that would be of interest to this audience.

1647a

The most detailed bird's-eye view of the burgh

This detail from one of the most famous early maps of Edinburgh shows the central part of the High Street or Royal Mile in the vicinity of St Giles, which had become a cathedral in 1633 when Edinburgh – formerly subject to St Andrews in ecclesiastical matters – was declared a diocese in its own right. The Parliament House, nestling in front of the cathedral, had only recently been completed, constructed at the request of Charles I in 1631–40 in the former kirkyard sloping down to the Cowgate. It was an impressive building; as Sir Robert Sibbald, Geographer Royal, described it in 1693: 'Here is one of the highest Houses in the World, mounting seven Stories above the Parliament-Court, and being built upon a great Descent of the Hill, the back Part of it is as far below it, so that from the Bottom to the Top, One Stair-Case ascends 14 Stories high.' To the left of St Giles is the New Tolbooth (marked 'ii'), constructed in the 1560s, and used variously as a market house, courthouse and gaol. Adjacent to it on the right are the multi-storey Luckenbooths, or 'locking-up shops', which formed quite an obstruction along the High Street, and were removed by 1817.

Further to the right is the High Cross or Market Cross (12), an impressive construction decorated with medallions and the town coat of arms. It had only recently been moved to this location from a site 45 yards further west, where it had been placed in 1617 for the visit of James VI and I. Far from being religious in character, market crosses commemorated the fact that the burgh in which they stood had been granted the right to engage in trade, a right granted by the Crown and not by the Church. In addition to being the usual site of public proclamations and announcements (local, national and commercial), Edinburgh Cross was the scene of an incredible variety of activities, particularly in the Jacobean period. At James VI's first royal entry into Edinburgh as an adult, it was host to an astrological display and figures of Peace, Plenty,

James Gordon, *Edinodunensis Tabulam* (1647)

EDINBURGH: MAPPING THE CITY

Justice and Religion, who addressed him in Scots, Latin, Greek and Hebrew. In 1599 the king allowed English actors to perform at the cross, and when this was preached against, the preaching was condemned by royal proclamation – again from the cross. The Gowrie brothers were hanged in chains here for conspiring to kidnap the king in the summer of 1600; and, perhaps most spectacularly, a 'juglar' called Robert Stewart, known as the 'Master of Activity', tied a tightrope from the cross to the steeple of St Giles and performed 'supple tricks' high in the air above a marvelling crowd. The current High Cross is a replica dating from 1885; the original was demolished in 1756, its functions having been largely superseded by the newspaper press.

A little further down the hill is the Tron (13), or weighbeam, a central part of Edinburgh's market trade, and beside it the Church of Christ the Saviour at the Tron (colloquially the Tron Kirk) (n), which was just nearing completion – the first seat rents were taken in the same year as this view.

1647

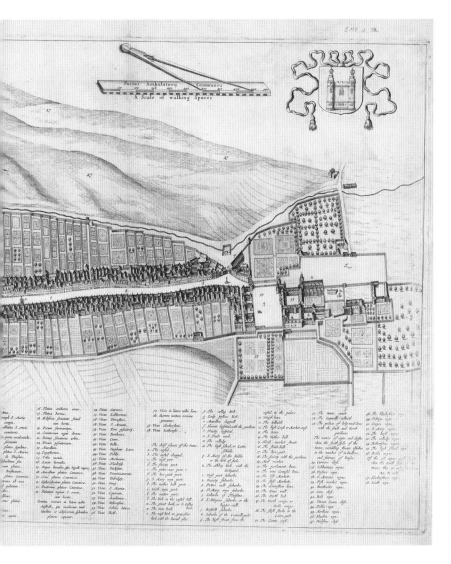

The numbers from 38 to 44 name all the wynds and vennels at right angles to the High Street and running alongside the traditional burgage plots: St Monans Wynd (38), Fish Market Wynd (39), Borthwick Wynd (40), Cons Close (41), Bels Wynd (42), Steven Laws Close (43) and Peebles Wynd (44). Most of these can still be identified today, and Old Fishmarket Close, Borthwick Wynd and Bells Wynd still carry the same names.

James Gordon (1617–1686), parson of Rothiemay in Banffshire, was undoubtedly one of Scotland's most skilled cartographers, drafting significant maps of towns and counties and assisting his father, Robert Gordon of Straloch, in completing the Blaeu Atlas of Scotland (1654). Edinburgh Town Council commissioned James to visit Edinburgh in 1647, in a brief interval of peace between the first and second Civil Wars, which as in England divided the population into royalist and 'puritan' factions. Also as in England, the puritans took and held the capital more or less from the outset, but remained less welcome in northern, western and upland regions. As a once-and-future Episcopalian and suspected anti-Covenanter, frequently rebuked by the Church for his lack of gravity, Gordon was an odd choice indeed; his winning the contract to produce this view may reflect that the uniqueness of his professional abilities transcended the vengeful religious politics of his age. Gordon's imaginary viewpoint and deliberately widened High Street flatter and promote Edinburgh, allowing us to see down all the vennels and to grasp the appearance and layout of minor as well as major buildings, which for the first time in Edinburgh's (or Scotland's) history are demonstrably realistic in both form and size. The Town Council were delighted and paid Gordon 500 merks for his labours (the then-equivalent of £27 15s 7d sterling, or about £57,000 now); they also elected him a burgess and guild brother as a token of their appreciation. The standard map or view of Edinburgh for nearly a century, often copied and reprinted, it is still popular today.

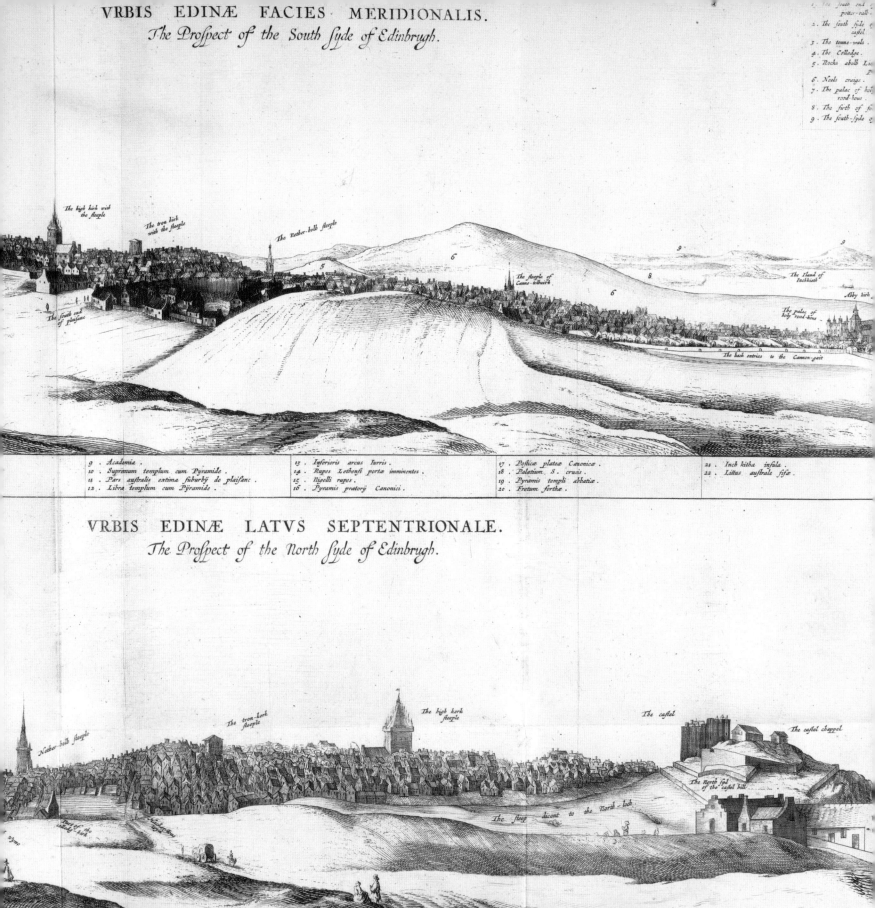

1647b

Viewing the Old Town from north and south

These little-known views by Rev. James Gordon complement his famous bird's-eye view of Edinburgh from the south [1647a]. His viewpoints – perhaps near what would become Nicolson Street on the south side, and what would become St Andrew Square to the north – give a uniquely important impression of these areas prior to later levelling and reconstruction. The shift in perspective to ground level also allows a clearer depiction of a number of features. The 24-foot-high town walls built after the disastrous Battle of Flodden (1513) had been extended in 1628–36 to enclose the splendid George Heriot's Hospital (then new) and Greyfriars Land, and would define the main boundaries of the burgh for the next two centuries. Houses mostly are shown with the tall, steeply pitched roofs that were at the height of fashion in the seventeenth century, though 'Scottish baronial' features such as crow-steps and bartizans are notable chiefly for their absence. The foreground figures, meanwhile, give good approximations of wheeled vehicles as well as male and female clothes. Six main gates or ports provided primary thoroughfares into the burgh, including the Bristo Port, which can be seen between Heriot's and the 'South end of plaisanc', whilst Leith Wynd port, by Trinity College Kirk, is labelled on the lower view. The walls turned north at this point to the Netherbow Port, and within the burgh of Canongate to the right are smaller gates through the 'heid dykes' at the end of tofts (burgage plots), providing 'back entries to the Cannon-gait' near Holyrood.

On the view from the south (top), Heriot's and Greyfriars are shown clearly to the left of Edinburgh Castle, and, proceeding to the right, St Giles, the Tron Kirk, the Netherbow Port, the steeple of the Canongate Tolbooth ('Canno-tolbuith'), and Holyrood Palace and Abbey. The castle still appears quite spartan, reminding us of just how much visible

---

James Gordon, *Urbis Edinae facies meridionales = The Prospect of the South Syde of Edinburgh. Edinae latus septentrionale = The Prospect of the North Syde of Edinburgh*, from *La Galerie Agréable du Monde* (1729)

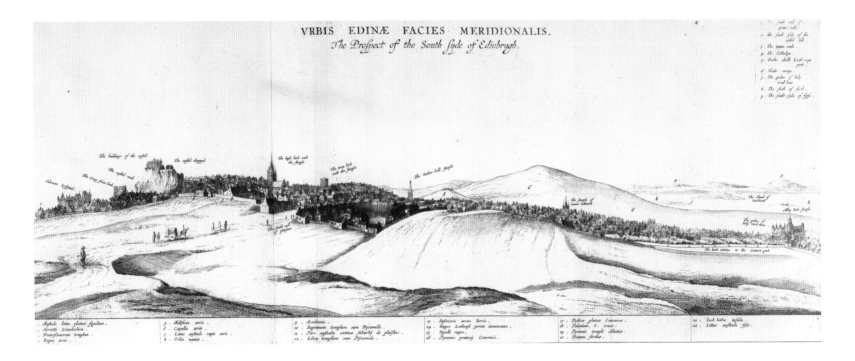

St Margaret's Chapel was still quite separate from the main castle buildings at this time.

on the site today is the product of its post-1660 status as a garrison for Scotland's first standing army. That being said, the castle would be occupied from as early as December 1650 by Oliver Cromwell's troops, following the Covenanters' noisy declaration of support of Charles II's claim to his executed father's throne, and their disastrous defeat by the Republican English at the Battle of Dunbar.

Compared with Gordon's bird's-eye view, the city's rocky, difficult and hilly topography is emphasised, though the Nor' Loch is in such a deep gully that it is not visible, only described. Behind the town to the north is 'The North Craigs or Neels Craigs': an earlier name for Calton Hill, after John McNeill, the clerk of Canongate, who purchased the area in 1588. The view across the Forth looks much the same today, with the island of Inchkeith and Fife both named in the key. The large building to the right on the view from the south is probably Multrie's Hill, a house and estate dating back to the acquisition of the area by Robert Multere in the fourteenth century. From the north, the familiar profile of Arthur's Seat and Salisbury Crags can easily be seen.

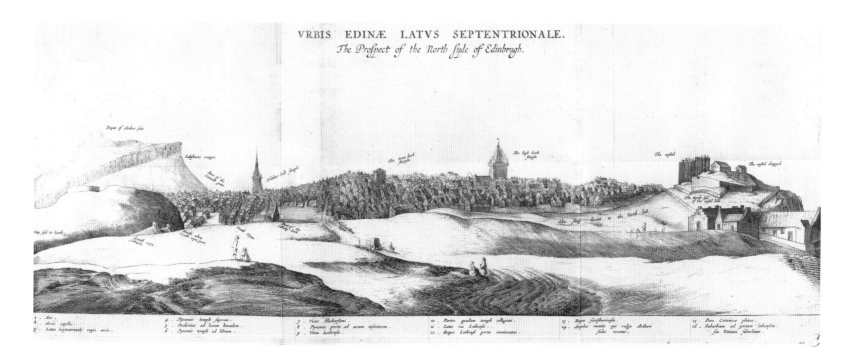

The plan and views were intended for inclusion in Blaeu's Atlas of Scotland, then nearing completion, and David Buchanan wrote a description of Edinburgh to accompany them; but in the event, neither text nor maps were ever included in it. The difficulties of communication between Scotland and the Low Countries occasioned by the 1650–60 Cromwellian military occupation of Scotland and the Anglo-Dutch War of 1652–54 coincided with the critical final years of bringing the Blaeu Atlas into print. Both the famous bird's-eye plan and the two views reproduced here were printed by Frederick de Wit in his town atlas of Europe in the 1690s, and subsequently by Pieter van der Aa in his *La Galerie Agréable du Monde* in 1729.

Surviving all the storms of Covenanting revolution, war and English occupation, Rev. Gordon remained minister of Rothiemay after the restoration of the monarchy and episcopacy in the early 1660s, and until his death a quarter-century later, at the age of 68 or 69.

St Giles and the Tron churches, with the Pleasance in the foreground.

granton
Lieth beacon
raustoun
Weirdy moor
new heaven
cytydale
Leith m.
Lieth
Laureston
murhous
Piltoun
E. Piltoun
Winnelstralee
Bonietoun
pilrige
wariston
mil
muttonhol
dryla
E dryla
crou
Inverlieth
Leith fl.
grotil
craiglieth
marsmil
canniemils
halkhil
restalrige
Loch end
grycrook
Dean
Waterlieth
brougton
abbyhil
raviston
mutvos
n Loch
W Kirk
cols
castle
cot Brig
derymil
delry
EDENBROUGH
damehead
wrightshous
Shyns
Saughtonhal
gorgie
bransfeild
didistoun
Loch
Priestfield
merchistoun
dalyetsmil
grang
Stenhousmil
Peppermil
powburn
camron
plewlands
Egypt
Bread craig
Inch
craigmiller
graysmil
Slatfoord
craighous
nedder
Liberton
ridehal
Bread
c. Lockhart
lille france
greenbank
edmondstoun
Liberton C.
Liberton K
over Liberton
Stenhous
colingtoun
colington mains
comeston
colington K
oxengang
guttors
Drum
fairmilhead
mortonhal
Gilmertoun

c.1682

A detailed manuscript map of Edinburgh's hinterland

Leith-born John Adair (1660–1718) is often thought of as the second major Scottish mapmaker after the pioneering Rev. Timothy Pont, who had surveyed nearly the whole of Scotland a century earlier [c.1610]. Although a number of other cartographers worked in Scotland in the seventeenth century – those involved with the Blaeu atlas including Robert and James Gordon, military engineers like John Slezer, and marine surveyors such as Greenvile Collins – Adair's work was the most extensive and included both county maps and sea charts. Adair was also perhaps the earliest Scottish mapmaker to use triangulation: we know that he spent money on compasses and telescopes, some purchased in the Low Countries, which would account for some of the improved distances and proportions in the present map compared with the earlier Pont/ Hondius map of the same area.

In May 1681, Adair was granted a licence by the Privy Council to produce maps of Scotland, and in the following year Sir Robert Sibbald (the Geographer Royal) commissioned him to produce maps for Sibbald's intended *Scotch Atlas*. Adair set to work on a number of county surveys at this time, including the county map of Midlothian from which this detail is taken.

This map of the broader Edinburgh environs superficially exhibits a continuity with the pattern of surrounding tower-houses, mills and routeways recorded by Pont a century earlier – but there are in fact many changes at a more specific level. For one thing, Adair provides us with much better detail of Edinburgh itself, in terms of the street layout, town walls and significant buildings. Moving from left to right, the castle, St Giles, the Tron Kirk, Netherbow and Canongate Kirk are all visible, as well as Heriot's Hospital in the foreground. Confusingly, most Scottish 'hospitals' at this date had no connection to medicine, being either private boys' schools (as here), glorified alms-houses, or some curious combination of the

John Adair, Map of Midlothian (c.1682)

two. In contrast to Heriot's and to Robert Gordon's Hospital in Aberdeen, the James VI Hospital in Perth was quite unusual in containing an 'infirmary to care for the sick' as well as the expected school and 'home for the poor'.

Adair's map is also the earliest cartographic record for many tower-houses: with Winnelstraelee, Nether and Over Quarryholes, Abbeyhill, Powburn, Bruntsfield, Whitehouse, Canaan, Egypt, Plewlands, Meggetland and Groathill all making their debut here. Some of these, such as Winnelstraelee, were recently constructed, but others were centuries older, and appeared now because of the larger scale of the map. There are two new bridges clearly shown across the Water of Leith, at Saughton and Colinton, as well as a number of new star-shaped mills. The village of Little France is reputed to have acquired its name owing to the size of the French entourage of Mary, Queen of Scots, who stayed at Craigmillar Castle, a short way to the north.

In 1633, the gruesome Burgh Muir gibbet [c.1610] was removed to avoid upsetting Charles I, who travelled by this route for his formal entry into Edinburgh, but we see it recorded in detail by Adair, now on the west side of Leith Walk. Adair also picks out the 'cytydale' (citadel) in Leith in some detail, as well as the Leith Beacon, guarding the Mussel Cape Rocks. The first Scottish lighthouse for the prevention of shipwrecks was constructed on the Isle of May, some 20 miles down the Forth, in 1635; earlier beacon chains were used to send important messages, usually warnings of invasion.

Adair's working life was not an easy one. He quarrelled badly with Sibbald over the rights to publish his maps, and was eventually freed from his contract with him in 1691. On the one hand, this lost Adair valuable time, leaving him without any engraved, published maps to show for his labours. On the other hand, the copies of his maps that he was required to supply to Sibbald are today his only manuscript county maps to survive, having entered the Advocates' Library with Sibbald's collections in 1723. Adair also struggled financially: official support through a tonnage levy on shipping from 1686 would never have covered his costs, even if he had not also had to compete head-to-head with John Slezer [c.1690] for the same meagre pot of funds. Adair appealed for subscriptions to support his county surveys, but the responses were always limited, and the ongoing lack of experienced engravers in Scotland at this time meant incurring the extra expense of commissioning Dutch or English engravers if Adair's maps were to be printed.

In 1688 Adair was living in Edinburgh 'in the Grassmarket near the muse well'. Nine years later he had moved again, acquiring a 'back land' on the south side of the Canongate. He died in Canongate on 15 May 1718, the only asset recorded in his will being £2 10s sterling owing to him for tonnage dues. Adair's maps were to suffer a limited afterlife: his wife, Jean Adair, handed them to the Exchequer Chambers in Parliament Square in 1723 in order to secure a pension, but all 39 items would perish in the Exchequer Office fire of 1811. However, in the 1720s and 1730s, the Edinburgh engraver Richard Cooper issued revised printed versions of some of Adair's county maps, including this one (right), with information updated to 1735.

Adair's precise drawing of streets and significant buildings provides particularly useful detail, given how few other maps were made at this time. Note the 1550s ramparts around Leith, and the Cromwellian citadel of 1656–57 to the west (see p. 47).

c.1682

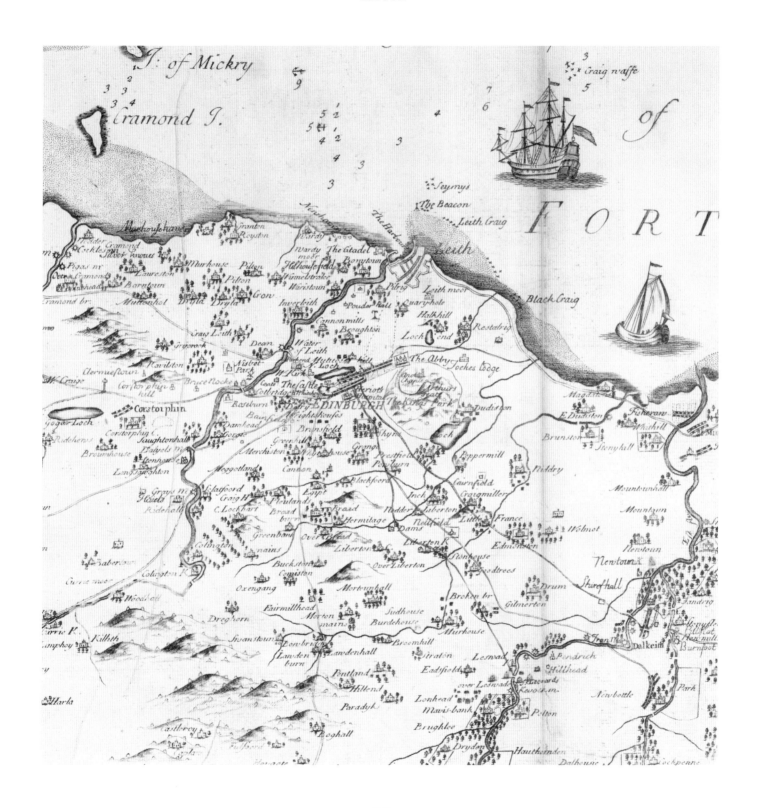

c.1690

The 'Queen Anne' view from Calton Hill

This striking and justly famous panoramic view of Edinburgh from Calton Hill compares well with James Gordon's view from the Lang Dykes in the 1640s [1647b], and Robert Barker's panorama of a century later [1790]. This is a deliberately flattering portrait, matching the accompanying description of Edinburgh, which 'far surpasseth all the other Cities of the Kingdom in the Stateliness of its Churches, the Beauty and Neatness of its publick and private Buildings, the Pleasantness of its Site, the Largeness of its Precincts, the Number and Opulency of its Inhabitants, and Dignity of its Rulers'. Whilst carefully drawn, it takes deliberate artistic liberties to create a more pleasing record of the town, adjusting perspectives and heights, and excluding the unsightly, to create an urban image of opulence and order. In the less biased assessment of a historian writing in 1999, the real Edinburgh was 'by even the most optimistic standards, disgusting. Overcrowding, disease, and squalor had given Auld Reekie its name and reputation.' Notable here for the first time are houses with multiple tall square chimneys rising directly from the wallhead: a style popular only in the seventeenth century, but which can still be seen on the east side of Pinkie House in Musselburgh.

In the foreground, the L-shaped building to the left of Trinity Church was Trinity Hospital, where, as Sir Robert Sibbald wrote, 'the poorer sort of Inhabitants, both Men and Women, are maintained splendidly enough, and have their own proper Chaplain'. The cross-shaped garden just beyond Trinity Hospital was the Physick Garden, containing 2,000 non-native plant species, which Sibbald and his distant cousin (but close friend) Sir Andrew Balfour, Bt, had established near Holyrood Abbey in 1671. It moved to the site shown here in 1676. The garden was used for the teaching of medical botany and supplied fresh plants for medicinal purposes; it was probably modelled on the gardens of the Duc d'Orleans at

---

John Slezer, *The North Prospect of the City of Edenburgh* (c.1710)

Blois in France, where Balfour had prudently sat out much of the puritan decade of the 1650s. Sibbald, though too young to have fought in the Civil Wars himself, witnessed the Cromwellian army's bloodthirsty sack of Dundee as an eight- or nine-year-old boy. Both founders earned their medical degrees in France, too, albeit at separate institutions. In keeping with the overwhelming majority of seventeenth-century British scientists, both men were committed royalists; Sibbald became physician-in-ordinary to King Charles II in 1682, and Balfour, president of the Royal College of Physicians three years later. Just behind the Physick Garden on the rising slopes was the Flesh Market, recently moved to this site from the High Street, just below the Netherbow Port. In 1763 the garden would move to the western side of Leith Walk, and in the 1820s to its current site by Inverleith Row.

John Slezer was an army officer and topographical draughtsman from an unknown German-speaking area of Europe, who first visited Scotland in 1669. His expertise as a military draughtsman secured him the office of Chief Engineer in Scotland from 1671, an appointment facilitated by his acquaintance with several of the nobility, including the earls of Kincardine and Argyll. Far from being a creature of the

c.1690

Restoration regime, however, Slezer would continue in crown service through the Williamite revolution of 1688–89 and the Hanoverian succession in 1714, serving six British monarchs of four different religious denominations. One of Slezer's official duties was to survey the defences of Scotland, and beginning around 1678 he worked steadily on views of castles, abbeys, towns and country houses; these were eventually published in 1693 as the *Theatrum Scotiae*.

The historical interest and artistry of Slezer's views in *Theatrum Scotiae* were more appreciated in later centuries than in his time. Slezer became heavily indebted as a result of poor sales of the volume, and even though a special tax on exports was imposed on his behalf, the receipts from this were shared with the surveyor John Adair and never amounted to much. By 1705 Slezer had moved into the debtors' sanctuary in Holyrood, where he remained until his death in 1717. Sibbald, in his capacity as Geographer Royal, wrote accompanying texts for *Theatrum Scotiae*, including quite an extensive description of Edinburgh. Slezer translated Sibbald's Latin text into English and published it without Sibbald's agreement and without acknowledging him as the author, which understandably caused a rift between them.

1693

Greenvile Collins and the professionalisation of chartmaking

Though liberally decorated with cherubs, which were the dominant motif in the English decorative arts of the Restoration and William and Mary periods (1660–94), this map is at its heart a businesslike tool for naval captains and merchant skippers hoping to enter the port of Leith in safety. This audience is implied by the three ships – a Scottish 22-gun warship, a Scottish unarmed merchantman, and an English 14-gun sloop – shown in the prospect at upper right, against the backdrop of Leith's tall, plain, geometrical and on the whole curiously modern-looking wharfside warehouses. The Scottish ships are marked by the Scottish red ensign, which carries the St Andrew rather than the Union in the canton. As well as the St George cross in her stern, the Englishness of the English naval ship is signified by the fact that, on the Union Jack in her bows, the St George cross covers the St Andrew, rather than the other way around. Scotland maintained its own navy, as well as its own standing army, until 1707.

Mapmaker Greenvile Collins rose to prominence through the personal intervention of King Charles II, one of the leading patrons of scientific enquiry and practical technological improvement in his own time, and indeed in any era of British history. Shipwrecked in northern Russia 500 miles northeast of Murmansk whilst trying to find a northeast passage to China – a goal of English mariners since the days of Edward VI (r. 1547–53) – Collins wrote a journal of his exploits that pleased the king, who gave him command of five Royal Navy vessels in succession. When Collins complained in 1680 of the poor quality of all available sea charts of the waters around Britain, most of which were both outdated and of Dutch origin, he was given command of the survey yacht *Merlin* and ordered to

Greenvile Collins, Map of Leith from the
north, with a prospect of the town from the east,
from *Great Britain's Coasting Pilot* (1693)

correct the problem himself. Two years later, in 1683, he was awarded the title of Hydrographer to the King, which apparently (given this map's publication date of 1693) was not revoked either by James VII or William and Mary. In any event, the massive cost of Collins's survey work – £1,914 sterling, or about £3.9 million today – was not funded by the Crown but by Trinity House, the private corporation which has had overall responsibility for England's lighthouses since 1514.

About one-third of Collins's charts of England and Scotland were privately printed as *Great Britain's Coasting Pilot* by Freeman Collins, possibly the cartographer's brother. Ironically, given their maker's strongly expressed (and in modern terms, correct) view that the coasts of Britain should be re-surveyed and re-drawn every few years, his charts proved extremely popular, and were re-issued – *without* substantive changes – 20 times between 1723 and 1792.

The sixteenth-century defences of the town of Leith, though evidently still fairly strong, are sketchy on the

A 22-gun warship of the Royal Scots Navy, against a backdrop of wharfside warehouses by Leith harbour mouth.

landward side: reflecting a re-focusing of concern towards foreign raids and invasions rather than civil conflict. Unsurprisingly, more attention is given to the shapes and positions of the 'Low water Mark', 'Becon' and 'Peere'. The Cromwellian citadel, constructed at vast expense in 1656–57 'to keepe in awe the chief citty of this nation', can be clearly seen to the west of Leith. Unlike other Scottish Cromwellian forts at Ayr, Inverness and Perth, which were built on previously undeveloped land, the construction of Leith citadel destroyed the Old Hospice and Church of St Nicholas as well as the burial ground of North Leith Parish Church. It originally had five bastions, with accommodation for officers, a barracks and a gunpowder magazine within the ramparts. Following the Restoration, the citadel was presented to the Earl of Lauderdale as a reward for his support for the Crown during the Interregnum. Soon sold on, it was rapidly developed for housing, and only its east gate in Dock Street survives today.

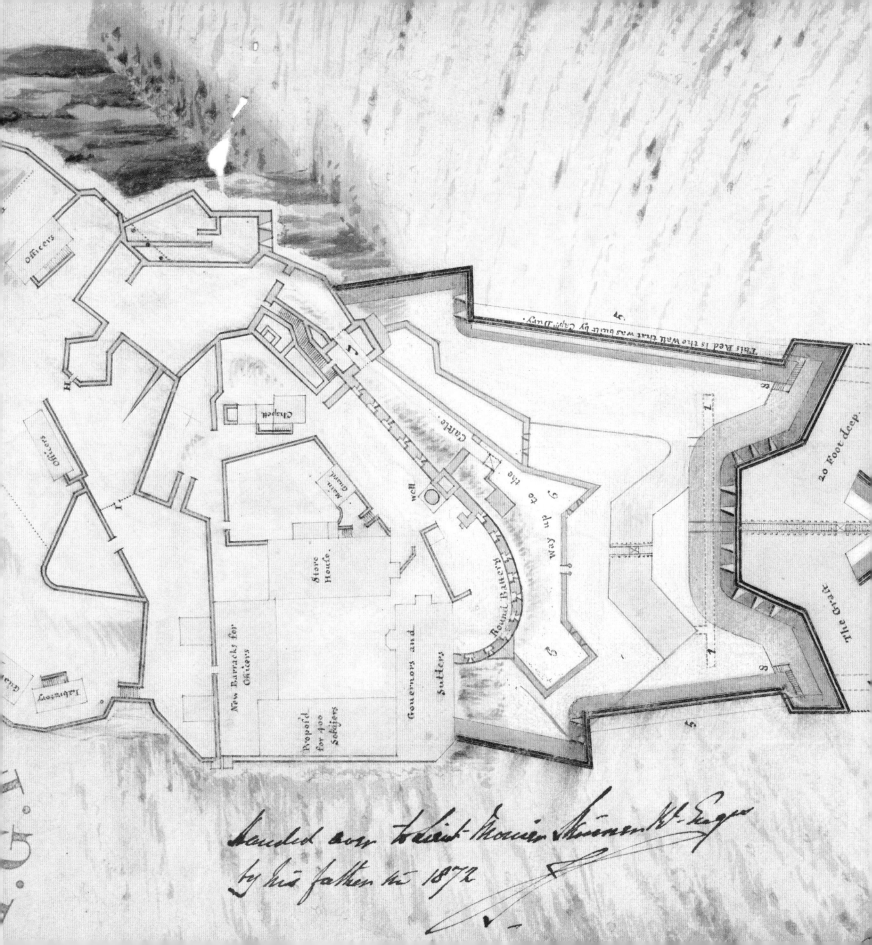

1710

'Le grand secret': re-fortifying Edinburgh Castle

As sharp-eyed readers will have surmised from the content of this book so far, military considerations underlay most mapping of Edinburgh in the sixteenth and seventeenth centuries, and this would continue to be the case through the first part of the eighteenth century as well. This striking hand-drawn plan of Edinburgh Castle, orientated with west at the top of the sheet, shows draft designs – never fully realised – for impressive external defensive works on its vulnerable side facing the town. It embodies the application in Scotland of the latest principles in military engineering, and in style, colouring and purpose it also reflects European military mapping conventions. From an overhead perspective, it shows existing walls and masonry in red, and projected defences in green and yellow. These illustrate the influence and principles of the greatest contemporary Continental engineers: France's Sébastien Le Prestre, Seigneur de Vauban (1633–1707) and Baron Menno van Coehoorn (1641–1704) of Holland. It is likely that only a handful of military officers would have seen this secret plan when it was new.

Following the parliamentary union of England and Scotland in 1707, the English Board of Ordnance reluctantly took responsibility for Scotland's castles and forts. Its traditional three-pronged 'broad arrow' symbol, really a pheon or heraldic arrowhead of the Tudor period when the Board originated, prominently announces the authorship and ownership of this document. (The 'I.G.F.' stands for 'Inspector General of Fortifications'.) Now largely forgotten, the Ordnance was a third branch of the British armed forces, which until 1855 had sweeping responsibilities for mapping, coastal defence, military engineering, munitions, and the design, manufacture and deployment of artillery. Its Master-General was theoretically equal in power to the Commander-in-Chief of the British Army, though often more powerful in practice: Wellington himself would hold the Master-Generalship for eight years,

---

Talbot Edwards, *A Plan of Edinburgh Castle* (1710)

after defeating Napoleon and immediately before becoming Prime Minister. The blue, not red, coats worn by British artillerymen and engineers from the earliest days of uniform reflected this separate heritage, and officers of these branches – in contrast to the army *per se* – could not buy and sell their ranks, but were promoted based on merit and seniority only. This was undoubtedly a reflection of the higher levels of specialist skill, including drawing and surveying, that were required of them.

Although the main threat to Scotland's castles in the decades that followed the Union would be from home-grown Jacobite rebels, they were also vulnerable to foreign powers, including the French, against whom British and allied forces were actively campaigning during the War of the Spanish Succession (1701–14). A French naval fleet had even been sighted in the Forth in 1708, and the Board were lucky that the Jacobite rebellion of that year – the second of five that occurred between 1688 and 1746 – fizzled out without major incident.

Owing to the vastly increased power and range that had been attained by heavy artillery by the late seventeenth century, traditional castle walls could be reduced to rubble all too easily, and the main defensive strategy involved trying to keep enemy guns as far away as possible. Edinburgh Castle was naturally strong on three sides, but its eastern flank had become steadily more vulnerable, and this was the focus of the new defences. From related plans, we know that Theodore Dury, Chief Engineer in Scotland, proposed an extensive 'hornwork' – a supplementary defensive structure comprising two demi-bastions joined by a curtain wall – across the area now forming the castle's Esplanade. Work began on this, termed 'le grand secret', in 1709; and a northern wall 250 feet long and 30 feet high is shown in red (that is, as constructed) on this plan of the following year. Nevertheless, there were clearly problems with Dury's proposal: not least, how near the hornwork's northern and southern walls were to the precipitous slopes on either side, and how close enemy artillery could still get to the castle's main gate.

Talbot Edwards, the Second Engineer, was therefore called in to propose a new plan. His hornwork was thinner and substantially longer, with a large 'ravelin' of outer slopes leading up to a 20-foot-deep ditch, which the defending army could access via small tunnels and galleries. At the base of the plan, again in red, are civilian buildings which Edwards recommended be purchased, and presumably demolished, to give better sightlines from the castle itself (an echo of Grange's demolitions of the 1570s [1577]).

For various reasons – particularly the parsimony of the Board of Ordnance in London, who struggled to find funds for construction work in Scotland – Edwards's ambitious plan was never executed. It is useful not only as a vision of what could have been, but also for the detail it provides of Edinburgh Castle at this date. For the first time, we can see a new wall of 1689 with its five large gunports linking the Half Moon Battery of the 1570s (marked 'Round Battery') to the Portcullis Tower, as well as the flimsier 'Countergarde' defensive walls below (6), where the Gatehouse visible today would be constructed in the 1880s.

Dury was responsible for constructing new battery walls – one to the west is still named after him – and the Queen Anne Building (shown here as 'New Barracks for Officers'). The plan also confirms that, at this time, the Great Hall was earmarked for conversion into a barracks for 400 soldiers. In 1730–37 the whole of the perimeter wall to the west and south was rebuilt, and in 1742 the dry ditch across the front of the castle was eventually completed; but it was not until the 1850s that the Esplanade as we know it was constructed in this area.

Edwards was formerly Chief Engineer of Barbados and the Leeward Islands, and adopted William Skinner [1750], who joined the Board of Ordnance in 1719 and went on to become Chief Engineer of Great Britain in 1757.

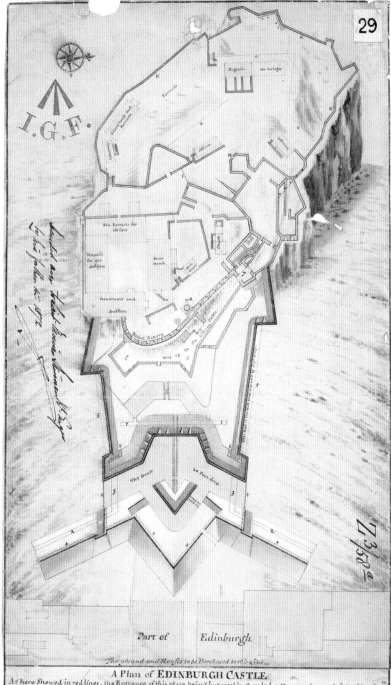

A Plan of EDINBURGH CASTLE

As here shewed in red lines, the Entrance of this place being but weakly guarded, a Hornwork was designed by Cap.ᵗⁿ Dury Anno 1709. and Carryed on upon the North wing, 250 foot in Length and 30 foot High — To this work Captain O Bryen made great Objections which being further Examined, the designe was found too broad for the Hill intended to be built upon, and too short to leave roome for standing of the old Countergarde to the Castle gate, or making a new one in the place of it as a Retrenchment within the Horne which would be much better. Wherfore to Correct these errors and save pulling down what was built by Capᵗ. Dury (though with some difficulty) This Hornwork as here express'd in strong black lines and green parapet round it was projected Anno 1710 — by Tallot Edwards her Majᵗˢ. second Engineer.

EXPLANATION

1. The double prickt line is a Gallery under ground for blowing up the Hornework when possest by an Enemy.
2. Another Gallery under the cover'd way for the same use, as also for clearing ye slope of ye Hill on ye North & South sides of the Horneworks by placing musqueteers there.
3. A Communication to these Gallerys through Coffers in the Graft.
4. The small black points are Pallizados in the cover'd way and Cross the graft which is 20 foot deep.
5. A Berme to be made of Masonry and earth for supporting the wings of the Hornework when the North and South sides of the Hill are scarped as proposed, that no attack may be made but on the front of this worke. 8. Are stairs to ye Berme
6. The Old Countergarde before the Entrance of the Castle.
7. The Portcullis gate. — 8 & 9. Retrenchments designed for better covering the west part of ye Castle where the ground is more then 20 and 30 foot higher then the walls below at K.

H A Gate to be taken down and removed to I.

A Scale of Feet

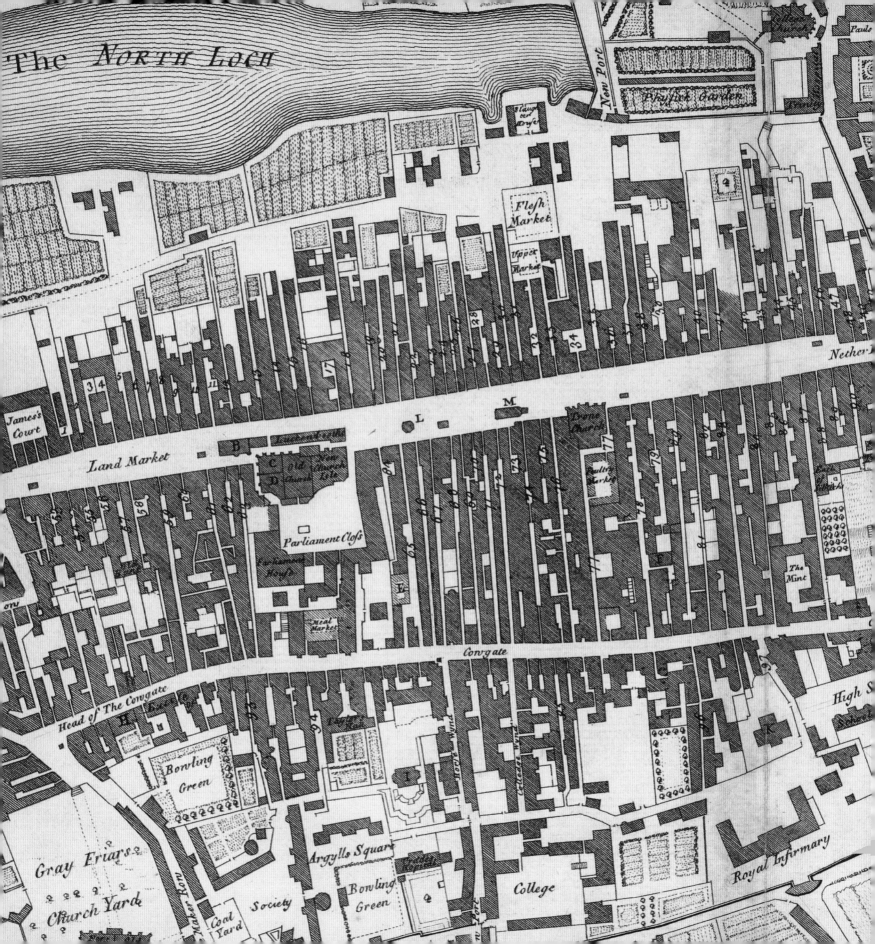

1742

The first overhead plan of Edinburgh

This map by William Edgar is famous as being the first directly overhead or 'ichnographic' map of Edinburgh, as distinct from either the county [c.1610] or the castle alone [1710]. The ready familiarity of this overhead perspective makes it easy to forget that, for many centuries, the bird's-eye perspective was predominant, and in fact continued to have its adherents, in later panoramas and views as well as in contemporary 45-degree web-mapping satellite views.

Edgar was born at Wedderlie in Berwickshire and apprenticed to an Edinburgh wright in 1717. By 1740, however, he was clearly working as a surveyor for the Board of Ordnance, and there are surviving maps by him of Peeblesshire (1741) and Stirlingshire (1743), and of waterways including the Forth and Don, as well as of military roads. Edgar is known to have travelled north with pro-Government forces during the final Jacobite rebellion of 1745–46, and to have died on 23 July 1746 at Fort Augustus, apparently of exhaustion.

We are fortunate that, in 1742–44, Edgar drafted this very detailed and accurate plan of Edinburgh, which was engraved in London by Paul Fourdrinier (1698–1758), child of French Protestant refugees from the repressive regime of Louis XIV (r. 1643–1715). The choice of Fourdrinier reflected Scotland's ongoing shortage of engravers as much as his skill.

Although slightly over-extended in a north–south dimension in its central area, the map is geodetically very accurate, perhaps reflecting Edgar's training in military engineering. In some ways, it is striking how little Edinburgh had changed in extent since the days of James Gordon [1647a, 1647b], with only modest expansion on the South Side and Portsburgh. However, within the town there had been some significant changes at a more detailed level, collectively representing the first stirrings of the grandly neo-classical city that would soon come to be. St Mary's Wynd Port and Leith Wynd Port have both been swept away, as part of the long decline in the usefulness – real and perceived

William Edgar, *The Plan of the City and Castle of Edinburgh* (1742)

– of civic fortifications. Forming one side of Argyle Square (laid out in the 1730s on land owned by the 2nd Duke of Argyll), we can see the Trades Maiden Hospital: a girls' school founded in 1704 by private banker and philanthropist Mary Erskine (1629–1707). The neo-classical Workhouse, with accommodation for 484 adults and 180 children, was begun in 1739 immediately to the south of Greyfriars, while the first Edinburgh Royal Infirmary – an immense, U-plan structure designed in the 1730s by William Adam in a partly baroque, partly neo-classical style – is clearly marked at lower right. Often hyperbolically and misleadingly described as 'Scotland's leading architect', Adam (1689–1748) would be more properly called the leading architect of his generation who worked primarily in Scotland. His Scottish contemporaries James Gibbs, Colen Campbell and James Smith were not only more successful architects on a pan-British basis, but arguably more talented; and Adam has been accused of lifting the non-baroque components of his Infirmary from a Board of Ordnance barrack-block. That being said, however, the vital contribution made by Adam and his architect sons Robert and James to the eventual look of the city of Edinburgh – as well as by the Board of Ordnance to British architecture as a whole – cannot be gainsaid.

Edgar also helpfully recorded 294 wynds, closes and vennels within the town walls and the Canongate, numbered on the map and listed in an accompanying table. He refers to the Lawnmarket as 'Land Market', its correct name at the time, reflecting the street's late-medieval and subsequent function as a marketplace for manufactured goods from the city's hinterland or 'landward', including cloth, stockings and thread.

The value of the map was clearly recognised by historian William Maitland (d. 1757), who was accused of using unfair methods to gain possession of it. Knowing or suspecting that the map was Edgar's sole asset, Maitland encouraged a local tailor to make a claim on Edgar's estate as a creditor, which led to the map's sale to Maitland in a public roup (auction) for £16. Although the Auditor of Excise as well as Edgar's sisters, Marion and Jean, subsequently accused Maitland of acting unfairly, he raised a counter-claim against them and proceeded to publish the map in his *History of Edinburgh* (1753).

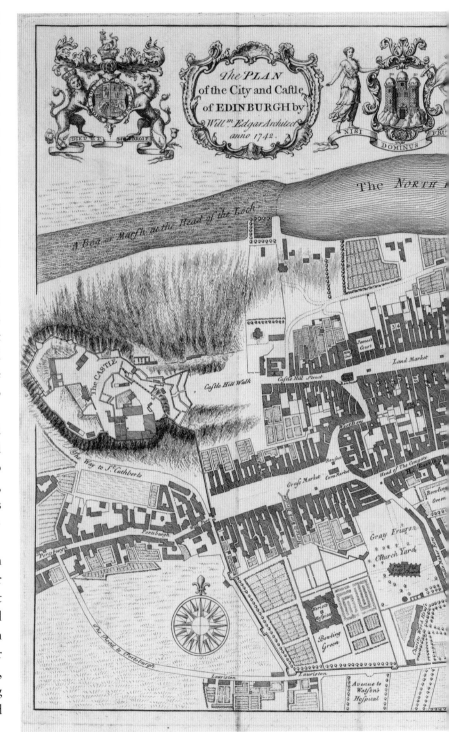

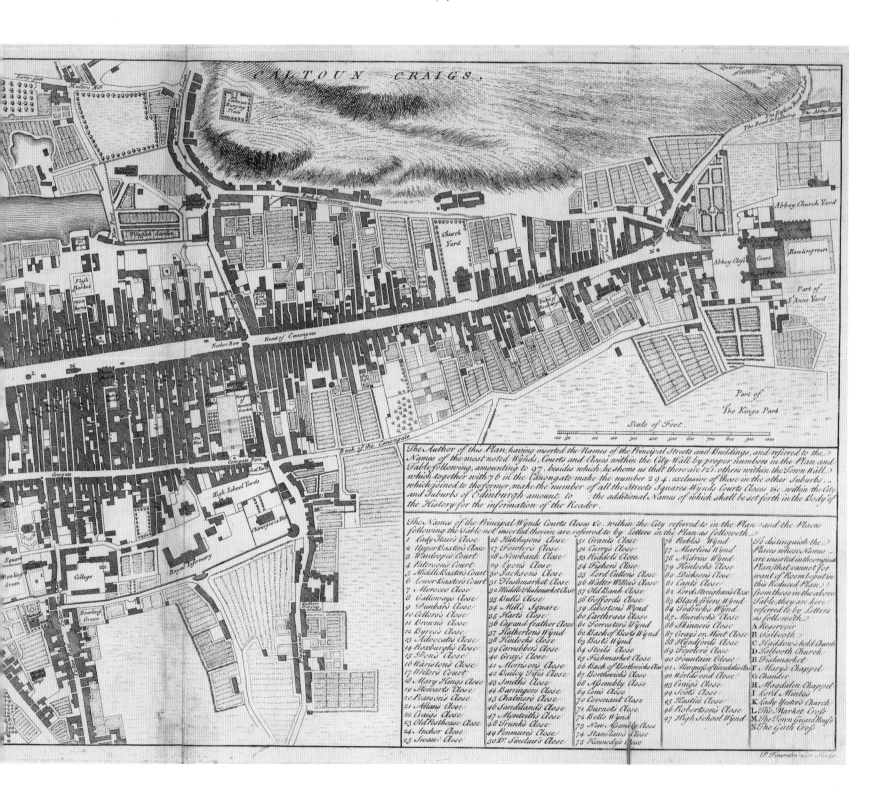

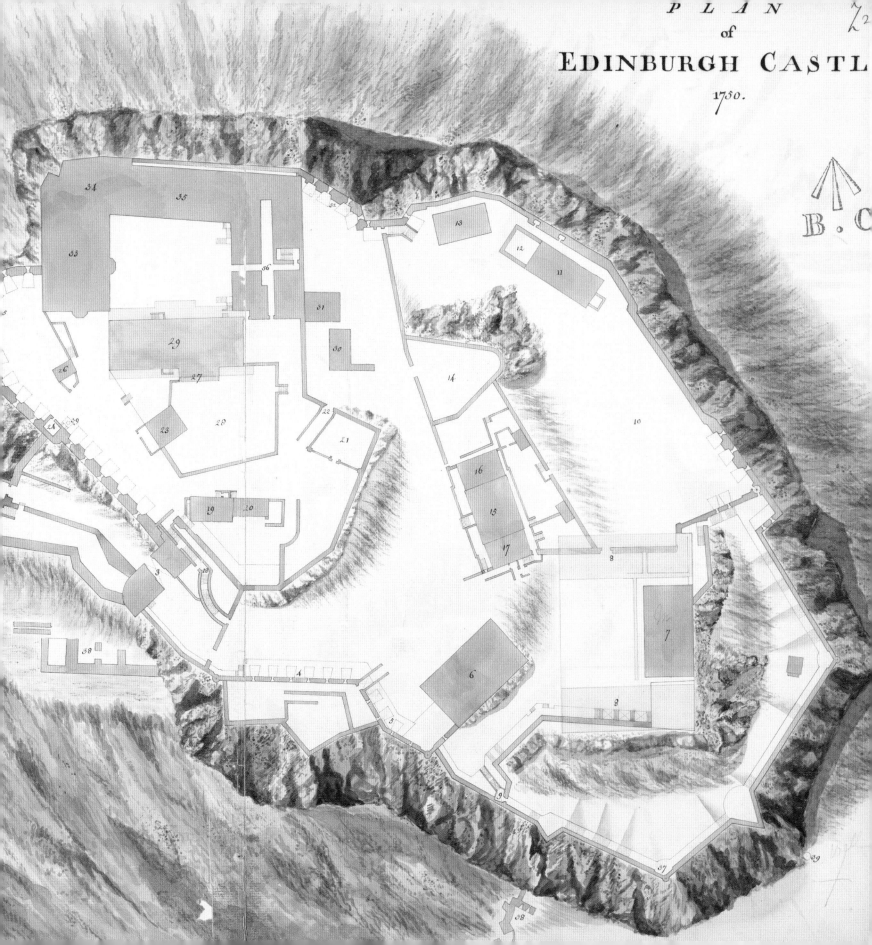

1750

Post-Culloden Edinburgh Castle as a base for the Roy Military Survey

This colourful map of Edinburgh Castle, with south at the top, was drawn by William Skinner (1699/1700–1780), who had been appointed Chief Engineer in Scotland from 1746 with orders to survey the works at Edinburgh, Stirling, Dumbarton and Blackness, as well as Fort William and Fort Augustus, following the great damage that these fortifications had sustained during the recently ended Jacobite rising. This map graphically depicts building work in the generation since the Queen Anne period (compare [1710]), as well as planned work: the buildings marked in yellow – an armoury/storehouse (8) and new barracks (29) – would not be constructed until 1753–55. Most interestingly, perhaps, the plan is contemporary with the drafting of the famous 'Great Map': the Military Survey of Scotland conducted from 1747 to 1755 under the superintendence of William Roy (1726–1790), then still a young civilian assistant to the Quartermaster-General's department, who with other engineers would have been based in the castle at this time. Obtaining his first military commission on completion of the Great Map in 1755, when he was not yet 30 years old, Roy would eventually rise to the rank of major-general.

Although Edinburgh Castle was nearly captured in 1715 by a Jacobite surprise attack through the sallyport on its western side, it had not really been put to the test, and in the final Jacobite rising of 1745–46 Prince Charles Edward Stuart had no heavy artillery that could have enabled a serious assault upon it. Optimistically and (in hindsight) tragicomically, each of the first four Jacobite rebellions, and indeed each of the Civil Wars of the 1630s–50s, was widely seen as the last thing of its kind that would ever occur in Britain. As such, construction work in the 1720s had been primarily directed to internal buildings, including a back barracks (11) adjoining a new malt kiln (12), and modifications to the entranceway and church porch. In spite of the grand Vauban-esque outworks planned

---

William Skinner, *Plan of Edinburgh Castle* (1750)

40 years earlier, the approach to the castle is shown clearly as a basic drawbridge across a dry ditch (1). The apparently uncontroversial presence of several areas marked 'ruins' (12, 38) may strike the modern reader as un-military, but we should remember that the cost of quarrying and shaping building stone was very high, and old stones – even from the Middle Ages – were reused wherever and whenever possible. It would also be hard to overstate the centrality to both military and civilian life of malt (partially germinated dried grain), the basic ingredient of both beer and whisky.

The neat neo-classical Governor's House (15), still extant, was built in 1742, and was flanked on either side with lodgings for two of the Governor's principal underlings, the Storekeeper (16) and Master Gunner (17). In 1748, a new powder magazine (7) was built; designed by Skinner, it was capable of holding more than a thousand barrels of gunpowder, and survived until 1897.

We do not know for certain where the famous Roy was based within the castle complex, but according to Chris Tabraham, Historic Environment Scotland's Principal Historian, there is considerable circumstantial evidence for this having been in the basement of the new Governor's House. The Military Survey had been encouraged, personally and financially, by David Watson, who from 1747 was Deputy Quartermaster General in Scotland, and would have lived in the Storekeeper's House. As civilians, both Roy and Paul Sandby – credited as being the principal draughtsman of the map – were unlikely to have been based in the buildings around what is now the Crown Square. The Storekeeper's House, moreover, was converted into an Ordnance Office by 1805. Roy and his engineers spent the summer months in active surveying work around Scotland – first in the Highlands until 1752, and then the Lowlands to 1755 – but spent the winter months in Edinburgh Castle drafting the map. In addition to being an important act of military cartographic surveillance, the result was the most complete and accurate map of mainland Scotland in the eighteenth century, and a formative influence on the modern Ordnance Survey (see [1752–55]).

Skinner was an acknowledged master of military engineering and draughtsmanship, and his plan – accurately scaled and carefully drawn, with conventional military colouring and standard aesthetics – illustrates these skills well. Although his main attentions by this time were already directed to the astonishing Fort George, a 42-acre site northeast of Inverness that was to preoccupy him for 20 years beginning in 1742, he was still keen to encourage vital improvements in the capital. Sandby, posthumously hailed as the 'father of English watercolour', was appointed Drawing Master at the Royal Military Academy, Woolwich, in 1768, from whence his style spread throughout the British world. He is considered a major influence on J.M.W. Turner.

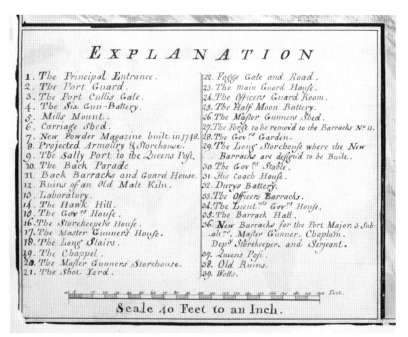

The table or legend identifying the buildings on the main plan of Edinburgh Castle (see p. 56). The projected armoury and storehouse (8) and the new barracks (29) were eventually constructed, but not until the mid 1750s.

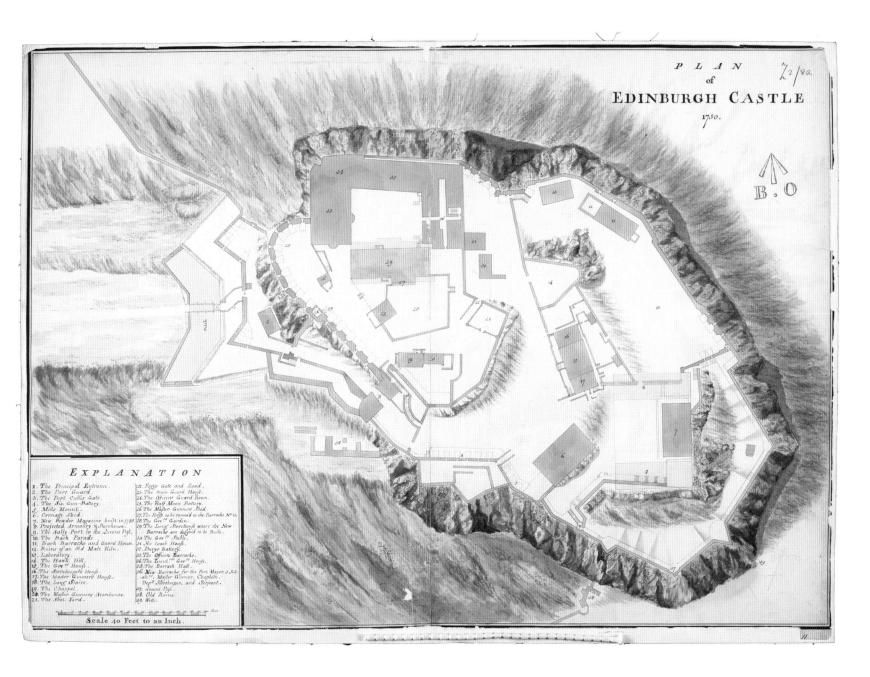

1752-55

The Roy Military Survey

The Roy Military Survey of Scotland rightly holds a very special place in the history of Scottish cartography. The story behind why the map was put together, its construction and draughtsmanship, and its influence on later maps and on the foundation of the Ordnance Survey are all especially important. It is also the most detailed map covering all of mainland Scotland in the eighteenth century, and therefore of irreplaceable value today for the history of landscape, architecture, agriculture, community development, transport and gardens.

The main impetus for this epic work was the Jacobite rebellion of 1745–46, when the House of Hanover's military commanders in Scotland found themselves 'greatly embarrassed for want of a proper Survey of the Country'. The Commander-in-Chief, William, Duke of Cumberland (sixth of eight children of King George II), had clearly recognised the importance of better mapping, and supported the idea; but the main funding and initiative for the work came from David Watson, the Quartermaster General in Scotland. Watson made the surprising, but in the event inspired, choice of putting the survey under the superintendence of 21-year-old civilian William Roy, the son of an estate factor from near Carluke in Lanarkshire.

It was a work of rapid reconnaissance and proceeded very swiftly: the Highlands were completed in five years, and the Lowlands in three, despite serious limitations in both manpower and resources. It was one of the earliest surveys in Scotland to use trigonometry, but only along selected traverses using fairly basic circumferentors (early cartographic tools that were later superseded by theodolites); many features were sketched in by eye. As Roy later remarked, it was more of a 'magnificent military sketch, than a very accurate map of the country' in which 'no geometrical exactness is to be expected, the sole object in view being, to shew remarkable things, or

William Roy, Military Survey of Scotland (1752–55). By permission of the British Library such as constitute the great outlines of the Country'.

Drawn in a pen-and-ink colour wash, its striking aesthetics illustrate emerging standards in military surveying and cartography, particularly through the choices of colour and selection of landscape features. The use of red for buildings and man-made structures, brown for roads of military value, blue-green for water, green for woodland, yellow for cultivated ground and buff for moorland was distinctively different from civilian mapmaking conventions. Paul Sandby (1725–1809) was described as 'the chief draughtsman of the plan', but he was assisted by Charles Tarrant, who later became a lieutenant-general, as well as by Roy himself, and (towards the end of the process) several others. The 'original protraction' for the Lowlands has an immediate and rough look to it, compared with the more finished and carefully executed 'fair copy' for the Highlands; but it is perhaps more faithful to the landscape, having undergone a less strenuous process of editing.

It is interesting and instructive to compare the Roy military survey to John Laurie's *A plan of Edinburgh and places adjacent*, which was drawn at a similar scale just a decade later [1766]. It is very unlikely that Laurie would have seen the Roy map, which was only shown to selected military personnel and the king for nearly half a century after its completion. If we ignore Laurie's unrealised proposals, such as his 'New Edinburgh', both maps show a reassuringly similar overview of the landscape: almost treeless and well cultivated, with a good scattering of significant tower-houses, estates and villages against a clearly recognisable network of routeways and tracks radiating out from Edinburgh. Even the main crossing points along the Water of Leith – of equal interest to civilians and military personnel – are comparable, as are the main roads' turnpikes and tolls, new developments of the eighteenth century. Both maps also show the recently drained Meadows, formerly the South Loch or Burgh Loch, an area actively reclaimed from the 1720s, particularly by Thomas Hope of Rankeillor, and often called Hope Park after him until the later nineteenth century. Middle Meadow Walk was opened in 1743, but the surrounding parallel grid of paths shown by Roy was not in fact implemented, and Laurie's work is perhaps closer to the reality on the ground. On the whole, the Roy Military Survey exhibits a greater selectivity with regard to place-names, even if the same buildings are depicted, as well as less interest in mills. However, Roy records several features around Arthur's Seat that Laurie omits, such as the Duke's Walk, St Anthony's Chapel and Salisbury Crags; and Roy clearly shows the gibbet near the top of Leith Walk, a significant landmark and of value in maintaining order and control, very much part of the Roy remit.

Although it had little impact upon later maps of Scotland, the Roy Military Survey was a seminal influence on British military mapping in the later eighteenth century, and, through Roy himself, on the early work of the Ordnance Survey. Roy had attained the rank of major-general by 1781, and regularly promoted the idea of a complete triangulation of Britain as a basis for better mapping. He led the English side of the geodetic connection of London and Paris by triangulation in the 1780s, begun by leading French cartographer and astronomer César-François Cassini de Thury (1714–1784), and went on to encourage the English mathematician and instrument-maker Jesse Ramsden (1735–1800) to construct a new and impressively accurate 36-inch theodolite. This was used to perform the primary triangulation of Britain in the 1790s, following the measurement of the Hounslow Heath baseline in 1791, laying the foundations for what would become the Ordnance Survey.

OPPOSITE. With spartan text, these details of Leith and Edinburgh capture many features of military interest, particularly patterns of buildings, streets, rivers/lochs, and densely settled country, crucial for planning attack or defence.

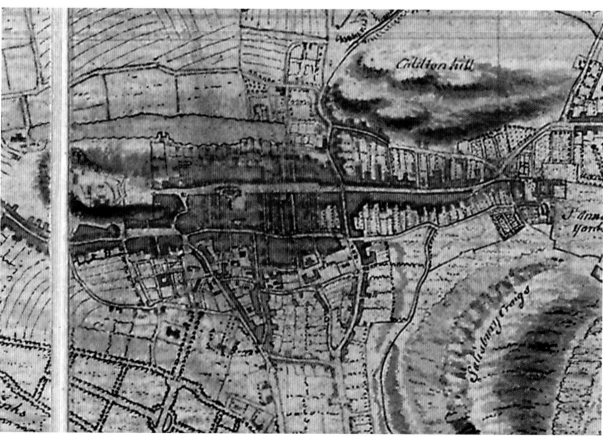

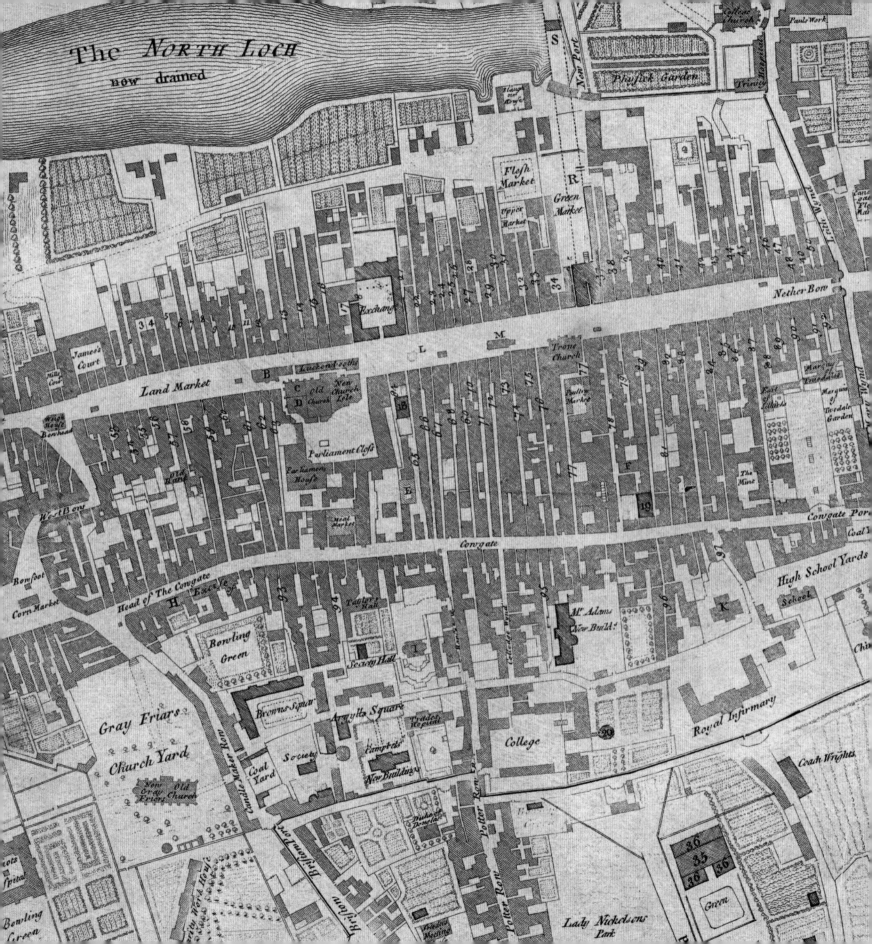

1765

New public and private buildings

It was always easier to correct an existing map engraved on copper than to create a completely new one, and this updated Edgar map of 1765 (compare [1742]) shows a number of new buildings with slightly bolder outlines, as well as demolished ones, often only partly erased.

In 1752 the Convention of Royal Burghs (CRB), a quasi-governmental advocacy group representing the commercial and legal interests of Scotland's most important towns, issued a highly influential pamphlet entitled *Proposals for carrying on certain Public Works in the City of Edinburgh*. In addition to reiterating the pressing need for new buildings, this advocated the construction of new streets on either side of the High Street, as well as the draining of the Nor' Loch, which, though once an important part of the town's late-medieval defences, had by the eighteenth century degenerated into a smelly and unloved municipal and agricultural waste pit. The CRB's proposals, which were accepted enthusiastically by senior figures in Edinburgh's civic administration, also specifically encouraged the construction of a building for merchants 'upon the ruins on the northside of the high street'. We can see the results of this in the newly constructed Royal Exchange buildings, completed in 1761 to a splendid neo-classical design by brothers Robert Adam (1728–1792) and John Adam (1721–1792). This complex would in the following century be developed and expanded as the City Chambers, the name by which it is still known. Faint numbers, only partly erased from the earlier printing's copper plate and still included in the key, mark the closes that were to be demolished or covered over by the Royal Exchange – including Mary King's Close (18), Stewart's Close (19) and Pearson's Close (20) – which together with Allen's Close comprise 'the town under the ground', a significant tourist attraction since 2003.

Further down the High Street, an area marked by pecked lines and the letter R indicates the intended course of the first

---

William Edgar, *The Plan of the City and Castle of Edinburgh* (1765)

ABOVE. Edgar's title cartouche, flanked by the Hanoverian version of the Royal Arms of Great Britain (with the famous white horse of Hanover appearing in the lower right of the shield) and the Arms of the City. At this time, Edgar's description of himself as an 'Architect' reflected his expertise in surveying, military engineering, and draughtsmanship.

OPPOSITE. Edgar's useful listing of all the vennels and wynds off the High Street updates Gordon's list from a century earlier, and extends beyond its box to include additions from Edgar's earlier 1742 map.

North Bridge (replaced by the current structure in 1896). The foundation stone for William Mylne's three-arched stone bridge had been laid in 1763, but there were a number of difficulties surrounding its construction – not least a partial collapse in 1769 – and in all it took nearly a decade to complete. Edgar's map also shows the former Hart's Close (35) and Cap and Feather Close (36) that were obliterated under North Bridge, while the Nor' Loch, although very much visible on the map, is noted as 'now drained'. In fact, drainage of the western end of the loch was not completed until 1820, at which date a submerged box was found to contain the skeletons of a man and his sister, executed by drowning in the loch for the crime of incest nearly 200 years earlier.

The two decades that had passed since Edgar's earlier plan saw an even greater number of changes on the South Side. To the south of the Bowling Green, Brown's Square has been laid out by James Brown. Further south, 'Campbels New Buildings' have been added to Argyle Square, and the corner bastion in the town wall south of Bristo Port has been erased. Further east, Lady Nicolson planned a new road, leading directly to her house and forking on either side of it. She also constructed a 25-foot-high corinthian column as a memorial to her husband, Sir James Nicolson, who had acquired the lands in question in 1727. The New Road ran adjacent to the new 'Riding School (35)' and 'Stables (36)', opened in 1764 as the Royal Academy for Teaching Exercises. In 1788, both Nicolson House and the memorial column were cleared away, so that the street would align better with the new South Bridge, completed in that year. The Riding School survived until the construction of the new Surgeons' Hall on the same site in 1829–32. To the east of the Pleasance, we can also see the new Bell's Brewery, which would occupy this site until the 1940s.

The Author of this Plan, having inserted the Names of the Principal Streets and Buildings, and referred to the Names of the most noted Wynds, Courts and Closes within the City Wall by proper numbers in the Plan and Table following, amounting to 97. besides which, he shews us that there are 121. others within the Town Wall which together with 76 in the Canongate make the number 294. exclusive of those in the other Suburbs, which joined to the former, make the number of all the Streets Squares Wynds Courts Closes &c. within the City and Suburbs of Edinburgh amount to 329. the additional Names of which shall be set forth in the Body of the History for the information of the Reader.

The Names of the Principal Wynds Courts Closes &c. within the City referred to in the Plan and the Places following the Table not inserted therein are referred to by Letters in the Plan as followeth

1 Lady Stair's Close	26 Hutcheson's Close	51 Grants Close	76 Peeble's Wynd	To distinguish the Places whose Names are inserted in the original Plan, that cannot for want of Room be put in this Reduced Plan from those in the above Table, they are here referred to by Letters as followeth
2 Upper Baxter's Close	27 Fowler's Close	52 Curry's Close	77 Martins Wynd	
3 Wardrope's Court	28 Newbank Close	53 Riddels Close	78 Nidries Wynd	
4 Paterson's Court	29 Lyons Close	54 Fisher's Close	79 Kinloch's Close	
5 Middle Baxter's Court	30 Jackson's Close	55 Lord Cullen's Close	80 Dickson's Close	
6 Lower Baxter's Court	31 Fishmarket Close	56 Walter Willes's Close	81 Cants Close	
7 Morocco Close	32 Middle Fishmarket Close	57 Old Bank Close	82 Lord Streighan's Close	
8 Galloways Close	33 Bull's Close	58 Goffords Close	83 Blackfriers Wynd	
9 Dunbar's Close	34 Mill's Square	59 Libertons Wynd	84 Todrick's Wynd	
10 Sellers's Close	35 Riding School	60 Carthraes Close	85 Murdoch's Close	
11 Browns Close	36 Stables	61 Forrester's Wynd	86 Skinner's Close	
12 Byres's Close	37 Hathertons Wynd	62 Back of Best's Wynd	87 Grayson Mint Close	A Reservoir
13 Advocate's Close	38 Kinloch's Close	63 Best's Wynd	88 Hyndfords Close	B Tolbooth
14 Roxburgh's Close	39 Carrubber's Close	64 Steils Close	89 Fowler's Close	C Haddon's hold Church
15 Don's Close	40 Gray's Close	65 Fishmarket Close	90 Fountain Close	D Tolbooth Church
16 Wariston's Close	41 Morrison's Close	66 Back of Borthwicks Close	91 Marquess of Tweedales Close	E Fishmarket
17 Writer's Court	42 Bailey Fifes Close	67 Borthwick's Close	92 Worlds end Close	F Mary's Chappel
18 The Royal Bank	43 Smith's Close	68 Assembly Close	93 Craig's Close	G Chander
19 St Cecilia's Hall	44 Barringers Close	69 Cons Close	94 Scots Close	H Magdalen Chappel
20 Monro's New Class	45 Chalmer's Close	70 Covenant Close	95 Hastie's Close	I Lord Minto's
21 Allans Close	46 Sandilands Close	71 Burnets Close	96 Robertson's Close	K Lady Yester's Church
22 Craig's Close	47 Menteiths Close	72 Bells Wynd	97 High School Wynd	L The Market Cross
23 Old Posthouse Close	48 Trunks Close	73 New Assembly Close	Q Sugar Work House	M The Town Guard House
24 Anchor Close	49 Penmures Close	74 Stanelan's Close	R The dott'd lines Shew ye Road along ye intended Bridge	N The Girth Cross
25 Swans Close	50 Dr Sinclair's Close	75 Kennedy's Close		O Alison's Square
				P Canongate Charity Work House

S Pier of ye Bridge Founded

1766

Fact, fiction and Craig's original plan for New Edinburgh

This map is deceptively believable as a record of the 1760s landscape around Edinburgh, but on closer scrutiny shows a number of projected and unrealised features, as well as a fashionable and attractive, but essentially stereotypical, rural landscape of fields and hedgerows that should not be taken too literally. While its depiction of the Old Town is bang up to date, with no Netherbow Port blocking the High Street (it had been removed as recently as 1764), it also anticipates a number of developments including George Square, which was only planned at this time. Likewise, the old Queensferry Road from the Grassmarket to Drumsheugh is shown attractively speckled with buildings, which seems implausible, given their absence from 1780s maps by Kincaid and Ainslie.

John Laurie was a local land surveyor, geographer and teacher of mathematics, and from 1763 he drafted a number of maps of Edinburgh and Edinburghshire (as the county of Midlothian was then called). These were engraved by Alexander Baillie (c.1710–1791), about whom little is known, other than that the fineness of the present work belies W.Y. Ottley's dismissal of him as 'of but moderate abilities'. (Given the engraver's name, and the fact that he lived in Canongate and Glasgow, Ottley's characterisation of him as an Englishman is likewise questionable.) Laurie had worked with the Town Council, and amongst other commissions produced site plans for the original North Bridge, which was completed at long last in 1772.

One of the most obvious features of this map is the striking outline for the New Town, named here as 'New Edinburgh', including a central square and bold diagonal streets forming a Union Jack shape. Though this flag-like plan was not executed, the New Town project was nevertheless the cultural apotheosis of Scottish unionism: the first adoption on a grand scale, within Scotland, of a generically British mode of Georgian neo-classical architecture which – though largely

---

John Laurie, *A Plan of Edinburgh and Places Adjacent* (1766)

the product of expatriate Scottish pens – would have looked equally at home in London, Dublin, Bath or Philadelphia. Interestingly, the first dwellings built in the New Town, in 1767, were a pair of semi-detached houses at 1–2 Thistle Court, the oldest extant (and presumably first) structures of their type in the country.

It has been argued on good evidence that this map represents leaked information that Laurie gained access to because of his connections in the Town Council. The architectural competition for the design of the New Town was advertised in April 1766, and seven entries had been received by the Architectural Committee by June. In August, they announced that James Craig's plan was the winner, albeit lacking 'so much merit as to be adopted', and expressing their hope that 'it might be of use in giving others hints to improve on'.

Laurie's plan illustrated here was published on 4 August 1766, and therefore seems likely to have been based on the winning plan submitted by Craig. But Craig's work was problematic: not least, in that the central square was where the crown of the New Town ridge was sharpest. Perhaps it was Craig who took the 'hints to improve on' from others, rather than the other way round: purging his initial plan of its weaker points, and replacing it with the world-famous adopted version, with its simpler gridiron outline and a square at either end. Craig was arguably not an architectural genius, and the final plan was described by A.J. Youngson, the main historian of the New Town, as 'poor in its simplicity, redeemed only by its superb site'. Laurie, for his part, hastily brought out a revised version of this plan in October 1766, showing the New Town broadly as it actually developed (p. 71).

Elsewhere, Laurie's plan shows a number of new buildings constructed since the Adair regional survey [c.1682]. Just south of Newhaven are the new estate houses of Laverockbank (acquired by the Leith merchant Robert Anderson in 1748) and Lilliput, probably named by Pierre de la Motte, dancing master in Edinburgh, after the fictional island in Jonathan Swift's *Gulliver's Travels*, published in 1726. The farm of Bangholm, just to the south, first appears on maps at this time too, and extending to the south is the earliest representation of what would become Inverleith Row in Canonmills. From medieval times, the bakers (Scots: baxters) of Canongate burgh were required to send their grain to Canonmills to have it ground, as this was the nearest mill that happened to be owned by the Augustinian canons of Holyrood Abbey, for whom both communities were named. The arrangement outlasted the Reformation, however, and even the sale of the Canonmills to Heriot's Hospital in 1637. This led to a lawsuit in the early eighteenth century between the baxters and brewers of the Canongate, on the one hand, and Heriot's on the other. The Court of Session agreed with the latter – that the brewers' secret operation of 'iron hand mills in their own houses' was 'an abuse' – but declined to impose any penalty.

Laurie went on to republish his maps in the following decades, this one in 1786 and 1811, with further updated and corrected details. In partnership with Robert Whitworth, he was also active in surveying the route for the Forth and Clyde Canal in the 1780s.

Royston
Broompark
Wardie
Newhaven
Ankerfield
Laverockbank
Trinity Mains
Royston Mains
Citydale
LEITH
Glassworks
Lilliput
Hillhousefield
San Mill
E. Pilton
Banghim
Wariston
Bonnyton
Brickfield
Hermitage
Bonnyton Mill
Pilrigg
Nether Quarry holes
Crue
Hawkhill
Inverleith
Canon Mills
Lochend
Craigleith
Upper Quarry holes
Silver Mills
Abby hill
Restalridge
Deanhaugh
Stockbridge
Caldton Hill
Broughton
Jock's Lodge
New Town
Caldton
Kings Park
Dean
Water of Leith
Coltbridge
EDINBURGH
Portsburgh
Colts
Duddingston
Castlebarns
Lauriston
Dalry
Fountain Bridge
Duddingston H
Hope Park
Gibbet Toll
Damhead
Toll
Wrights houses
Prestonfield
North Merchiston
Rosehall
Gorgie
Clearburn
Mr Warrenders
Merchiston
Greenhill
Whitehouse
Newington
Peppermill
Toll
Gorgies Muir
Mayfield
Redhouse
Grange
Cameron
Tipperlaw
Morningside
Pow burn
Craigmiller Bridge
Meggland
Miresyde
Blackford
Canaan
Briggs of Braid
Inch
Slateford
N. Liberton
Craiglockhart
Craighouse
Penlands
Mains
N. Craigs
Nether Braid
Hermitage
S. Mains
S. Craigs

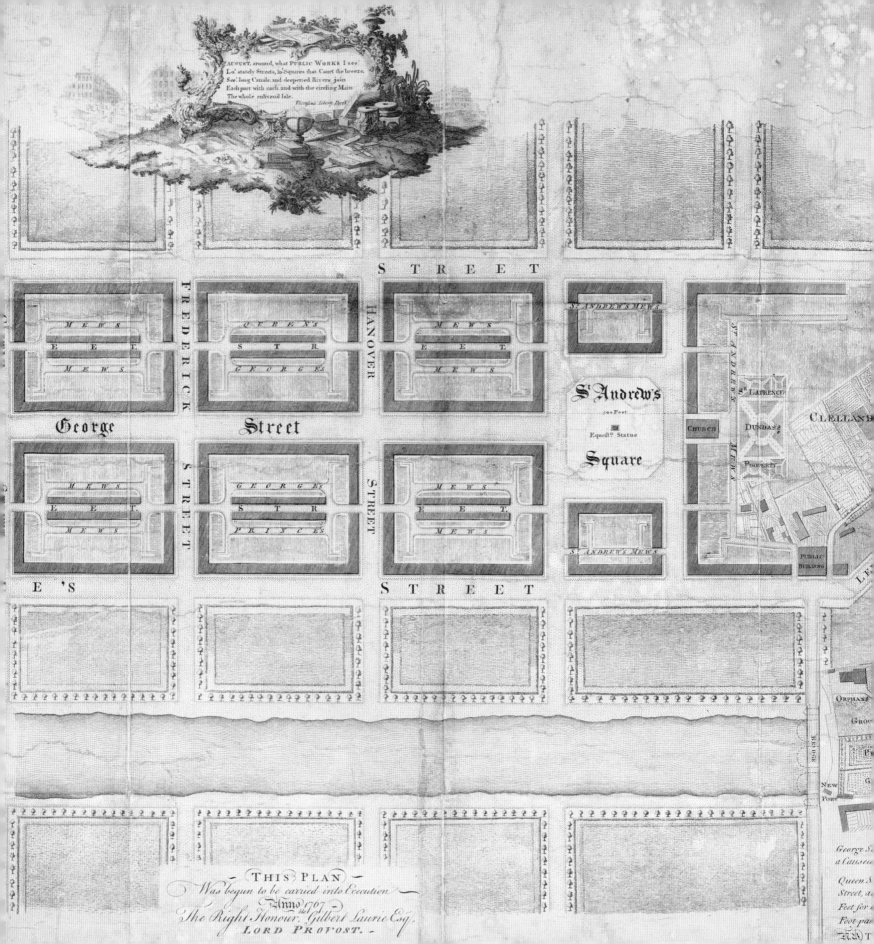

1767

James Craig's proposals for the New Town

This plan for the proposed New Town of Edinburgh made the 27-year-old James Craig (1739–1795) famous, for a time; but there were a number of other proposals for developing the area, both in Craig's day and earlier.

While he was Duke of York and *de facto* regent of Scotland in 1681, the future King James VII (r. 1685–88) had drawn up plans for the city's expansion to the north, and in 1687 followed this up by granting a northward extension to the royalty boundaries (the area enjoying the privileges of the royal burgh), along with a charter for improved roads and bridges. In a way, the entire New Town project makes more sense in the context of late-seventeenth-century schemes for the rebuilding of London after the Great Fire of 1666; and even the 'Union Jack' arrangement of the New Town envisioned by Laurie [1766] bears more than a passing resemblance to Sir Christopher Wren's unexecuted scheme for London (p. 75), at least as the latter was engraved in 1744 by Fourdrinier (see [1742]).

However, King James's Edinburgh plans were aborted in 1688 when he was forced to abdicate in the Glorious Revolution, a coup d'etat led by his Dutch nephew/son-in-law, William of Orange. This event was at the root of all the subsequent Jacobite rebellions – James was referred to as 'Jacobus' on the nation's coinage and official documents – as well as the long-standing political enmity between the Whigs (who continued the traditions of the Williamites) and the Tories (many of whom were Jacobite sympathisers). Ironically, perhaps, given that the deposed James was a Catholic convert and William was raised a Presbyterian, the latter took almost no positive interest in Scotland, where he is remembered chiefly for authorising the Massacre of Glencoe in 1692, and refusing to help establish a Scottish colony at Darien in Central America in 1698.

James Craig, . . . *Plan of the New Streets and Squares,*
Intended for [the] *Ancient Capital of North-Britain* . . . (1768)

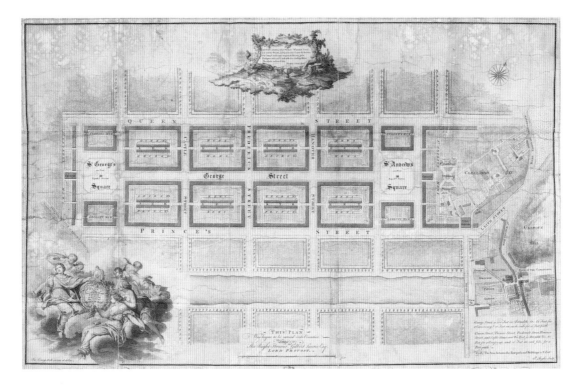

Curiously, in 1728 – many years after King William's death – the still-exiled Jacobite Earl of Mar drew up a plan for Edinburgh with strong similarities to Craig's: proposing a main street along the central ridge, the Lang Dykes [1773], and gardens sloping down to the north and south.

The finalised plan shown here was issued on 1 January 1768. Earlier surviving versions represent revisions to Craig's original plan of 6 June 1767, which is itself lost. Some of these revisions were the result of consultation with King George III, who suggested that St Giles Street – whose name reminded him of a part of London 'always infamous for its low and disorderly inhabitants' – should be renamed Princes Street. The princes in question were his two eldest sons, the future King George IV (r. 1820–30), and the grand old Duke of York who had ten thousand men (in the words of the old song, dating to a lacklustre military campaign of 1799). Queen Charlotte, meanwhile, felt that Queen Street would sound better than Charlotte Street. Her wishes were adhered to, but in the event the queen would have St George's Square named Charlotte Square after her as well, from 1785; this ruined the perfect unionist symmetry of saints' names in the plan shown here, but had the merit of avoiding confusion with the other George Square.

There are also later versions of Craig's plan, including one showing a circus at the junction of George Street and Frederick Street, although this was never built. In many ways, these alterations were mere tinkering with an essential form that remained unaltered throughout: a rational, ordered, symmetrical grid of streets that not only chimed with Enlightenment ideas in a graphical form, but also with the prevailing French doctrine that new towns should glorify the reigning monarch. Craig's dedication left no room for doubt about this last point, describing George III as 'The Munificent Patron of every polite and liberal art', Edinburgh as 'His ancient Capital of North Britain', and the New Town as 'the happy consequences of the peace, security and liberty his people enjoy under his mild and auspicious Government'.

Craig was also a huge admirer of his uncle, the poet James

1767

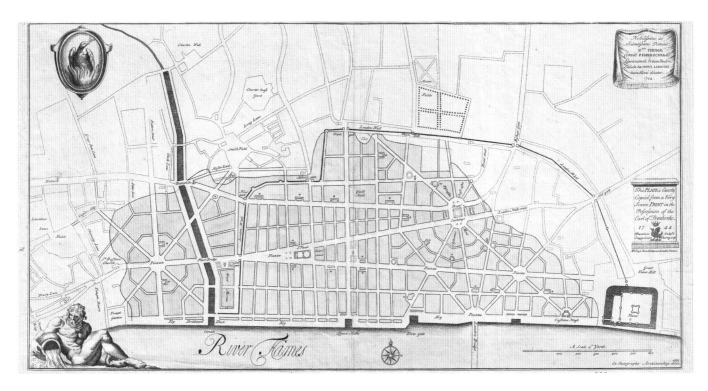

Thomson (1700–1748), who had written the words to 'Rule, Britannia!' in 1740. In this version of the New Town plan, Craig included a verse from Thomson's book-length poem *Liberty* (1735–36), which traced its titular concept from its origins in classical antiquity down to its achievement of perfection in, as Thomson saw it, the Williamite Revolution of 1688. The lines chosen for inclusion on the map were imbued with the sort of lofty sentiments that Craig might well have penned himself for the occasion:

> August around, what PUBLIC WORKS I see
> Lo stately Street, lo Squares that Court the breeze,
> See long Canals and deepened Rivers join
> Each part with each and with the circling Main
> The whole enliven'd Isle.

The foundation stone for the New Town was laid by Craig on 27 October 1767, and in the 1770s he went on to design the Physicians' Hall on George Street, as well as St James Square at the east end of Princes Street, though his 1785 plan for the development of the South Bridge area was not adopted.

Sir Laurence Dundas, the wealthy politician and landowner who had secured the Parliamentary Act that led to the building of the New Town, had also purchased in advance the land immediately to the east of St Andrew Square. Although Craig had intended that a church should be built on the site, as shown, Dundas had his own palatial villa (now the Royal Bank of Scotland) constructed there; St Andrew's Church had to be built at 13 George Street instead. Later plans of the developing New Town in the 1780s and 1790s show similar divergences from reality [1780].

This map was engraved by Patrick Begbie, who was originally from Edinburgh, but moved to London around 1774. By 15 June 1776, he was under lock and key in the King's Bench Prison for debt. Craig, too, became insolvent, and died unmarried, unmourned, and largely unremembered, with a negative net worth of £143. His Physicians' Hall was demolished in the 1840s, and most of St James Square followed suit in the 1960s.

75

HERIOTS HOSPITAL GROUND

Barron Ords Garden

Queens Street

Clellands Few

Barjargs Road to Stockbridge

Part of Andersons

Georges Street

St Andrews Square

Part of Woods

Farm

Register Office

Few

Farm

Long Dykes

St Andrews

Long Dykes

Princes Street

Play House

Methodist Meeting Ho.

ARE FOOTS PARKS

Orphan Hospital

Lady Glenorchys Church

Canal Street

College Church

Trinity Hospital

Intended Canal

Calton

from Canon Mills

Bridge

High St.

James's Court

Luckenbooms High St.

Land Market

The CASTLE

Castle Hill Walk

Castle Hill St.

Welbow

Parliam Closs

Roy Exc

Road from Queens Ferry

Cowgate

Lady Yesters Church

Cowgatehead

Graff Market

Adams Sqr

Portsburgh

Gardners

Browns Sqr

Royal Infirmary

Whitburn

Factory

Candlemakerrow

Argyles Sqr

Tradeshospital

College

HIGH RIGGS

Heriots Hospital

Gray Friars Churches

Bristow St.

Scoeders Meeting Ho.

Park of Relief

Charity Work Ho.

Police Row

Rideing School

Tiviot Row

Nicolhous

1773

The earliest buildings of the First New Town

In order to construct the New Town, Edinburgh Town Council had to engage in a series of acquisitions of land to the north of the Nor' Loch, and this map of 1773 shows some of these, along with the transitional patterns of roads and buildings. The rising ground between the Nor' Loch and the original Lang Dykes road (running between what would become Princes Street and Rose Street) had been acquired by the Hepburns of Bearford in 1645, who sold it to the Town Council in 1717; it was often known as Bearford's or (as here) Bare Foot's Park. However, the single largest landowner in the New Town area and immediately to the north of it was Heriot's Hospital, as indicated at upper left.

The outlines of Princes Street, George Street and Queen Street are all shown, and mapmaker Andrew Bell (1726–1809) also included a note with specific details of their plans:

The New Town is designed to consist of 3 principal Streets, viz. Queens Street, George Street, & Princes Street, 100 feet broad each, & two Mewse Streets betwixt them, 30 foot broad each, besides Canal Street, it is likewise to have other 5 cross streets 90 feet broad each, paralel to St. Andrew Street, & the other formed on the west side of the Square, the town to terminate on the west in a square, similar to that of St. Andrews Sq.

Canal Street was to run along the ornamental canal proposed to replace the Nor' Loch, but was only ever built upon at its far eastern end, and absorbed into Waverley Station from the 1850s.

Bell's plan also captures the earliest building development

Andrew Bell, *A Plan of the City of Edinburgh, with all the New Streets, Avenues, Buildings, Squares, Courts &c Within & Round the City, since 1741 till this Present Year* (1773)

at the eastern end of the New Town, including Sir Laurence Dundas's exceptionally elegant neo-classical villa on St Andrew Square, built 1772–74 to a plan by Sir William Chambers. After Dundas's death it would be used as an Excise Office [1804] – hence the royal arms still visible on the pediment – and, after 1825, as a bank. Further to the southeast we can see the outline of the planned Register House, where construction began in the following year, and the Theatre Royal or Play House, founded in 1768, around what would become Shakespeare Square a few years later. The vital North Bridge, providing the direct thoroughfare into the New Town from the south, had only just been completed. We can also see the final days of buildings that would be swept away in the 1780s for the construction of South Bridge, completed in 1788 just to the east of the Tron Church (marked 'H'). The wynds just south of the this, Peebles (74), Marlin's (75) and Niddry's (76), were all cleared as part of the same process. Also demolished at this time was the City Guard (72), a squat and ugly building described by Sir Walter Scott as a 'long black snail crawling up the middle of the High Street and deforming its beautiful esplanade'. Edinburgh's Town Guard, a paramilitary police force consisting of three companies, each with 29 officers and men, used this as their headquarters; it also contained a dungeon or 'black hole' for prisoners at its western end. Drastically reduced in numbers to 31 men in total, the Town Guard moved into New Assembly Close (70) until their disbanding in 1817. This important step in the evolution of modern British unarmed policing was enabled by the success of the Police Acts for Glasgow (1800) and Edinburgh (1805). It is a widespread, but quite erroneous, belief that such reforms sprang directly from the brain of Home Secretary Sir Robert Peel, Bt, who was in fact only 12 years old when the Glasgow act received the royal assent. That being said, Peel's relatively prompt application of Scottish Enlightenment ideas to England and Wales was perhaps radical enough to justify his enduring fame.

The map is also interesting for providing significant evidence of religious nonconformity, including an octagonal Methodist meeting house between the Play House and Calton Hill, and an 'English', i.e. Anglican, chapel in the Old Town. The existence of these buildings, as much as the engraver's willingness to portray them, probably owed something to the novel atmosphere of religious toleration that surrounded the young George III (r. 1760–1820). In part this was a pragmatic response to Britain's sudden and unexpected acquisition of 50,000 Catholic subjects with the capture of French Canada in 1763; but beginning in 1778, the king would go on to allow practising Catholics to inherit land, operate schools and serve in the British Army and the legal and medical professions, paving the way for the Relief Act of 1791. Similar relief for Scottish Episcopalians followed in 1792. Another detail of interest in the far west is St Cuthbert's Church, 'presently taking down to be rebuilt'. St Cuthbert's had been described as dangerous in 1772, and was reconstructed between 1773 and 1790 (and again in 1892–95). The octagonal Methodist chapel, built in 1764, was demolished in 1815 to make way for Regent Bridge.

Bell was an engraver and publisher whose premises at this time were located on the south side of Parliament Close, but which moved to Lauriston Lane from 1777. He was apprenticed to Richard Cooper until c.1748, and in turn, from 1782, trained both Hector Gavin and Daniel Lizars: two of Edinburgh's most successful engravers in the early nineteenth century. Bell was also one of the original partners in the *Encyclopaedia Britannica*, engraving plates for the first edition; he became its sole proprietor on the death of Colin McFarquhar, the printer, in 1793. Bell's attractive and detailed map illustrated here was dedicated to Gilbert Laurie (1729–1809), who was Lord Provost of Edinburgh in 1766–68 and again in 1772–74, perhaps in the hope that Laurie would lend his political or financial support to further commissions.

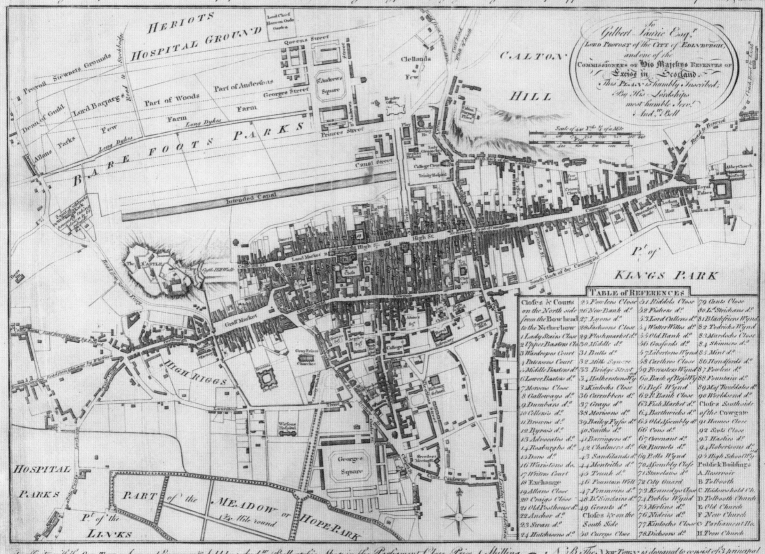

North Loch and other Grounds belonging to the Cit[y]

(Map labels:)
- vernor of the Castle
- Meadow
- Bark hill
- Ramsay's
- Tanner's Work
- GARDEN
- McCallum's
- Youls
- Mr Tods
- Reservoir
- Street
- Byres Cl[ose]
- W E (compass)

Scale: yards, 100 200 300

Left table:

	A:R:Fo:
Meadow twixt the Load... the Timber bridge... courses, also for the Tanwork Bark	5"2"34 4/10
...th Side ye Drain	—"—"20
...foot of Wilson... piece of Grass twixt the Tan bridge	0"0"2 4/10
	1"3"24 2/10
Rent	A: 7"3"23 3/10

Right table:

Measure of Garden Ground — Scotch

	A:R:Fo:
South Side the North Loch	
McCallum's, east side of the Tan Work, including old houses &c	2"2"—
Youls, east of D{o}, including an old claim at a	1"0"8
Ingles's twixt D{o}. & Hoggs 2 R: 2 Fa.	
D{o}. on both sides the Fod Road 1 R: 5 Fa.	
Ingles three parcells exclusive of ⅞ fallin in ye Road	0"3"7
	4"1"15
Hoggs Few twixt the two lott	0"0"33
Saded Yellow, Garden Ground	4"2"8
Shaded Green, Loch Ground	7"3"23 3/10
Totall for Rent	A: 12"1"31 3/10

1779

Land disputes around the Nor' Loch

This little-known hand-drawn map, based on a measured survey by John Lesslie, shows the competing claims to the newly drained land formerly occupied by, or adjacent to, the Nor' Loch. Although most title to land in earlier centuries had been described in prose, by the second half of the eighteenth century it was customary for maps to be used, based on a measured trigonometrical survey, especially where there were disputes over ownership. The period from the 1760s to the 1830s was the heyday of the private land surveyor, whose work proceeded hand-in-glove with agricultural improvement and reorganisation in a multiplicity of forms through maps, especially in rural Scotland.

The Town Council were well aware of the growing importance of the site, acquiring Bearford's Park in 1716 and actively feuing the eastern end by the North Bridge for building in the early 1770s. This was abruptly brought to a halt through legislation in 1776 (see [1819a]) which confirmed that the green meadow shown here, at least as far west as the Tan Yards and Hanover Street, would be 'kept and preserved in perpetuity as a pleasure ground'.

The eastern end of the loch had been partly drained in 1763 for the construction of North Bridge, shown on the far right-hand side of this map. However, it was not until 1790 that a drain was constructed from St Cuthbert's Church through the loch bed; west of the Mound, standing water remained until the 1820s. On the south side there were multiple owners and claims, with their precise locations clearly shown by reference to particular buildings or wynds off the High Street, including Byers Close and Mary King's Close. At the far west, the 'Meadow claimed by the Governor of the Castle, measuring 1 acre and 3 roods' (1.75 acres Scots or about 2.3 acres English) was described as disputed. The Governor, though

John Lesslie, *A Plan of the North Loch and Other Grounds Belonging to the City of Edinburgh* (1779)

acting on behalf of the Crown, evidently lost this patch, since later maps by Ainslie [1804] and Kirkwood [1817] confirm its ownership by the City of Edinburgh at those dates. Just to the east was Ramsay's Garden, extending down from the famous 'Goose Pie' house constructed by the poet Allan Ramsay in 1751. After he died in 1757, the house passed to his son, also Allan Ramsay, Principal Painter in Ordinary to George III. Above this garden we can also see the old Reservoir, which dated from the 1670s, and by the time of this map supplied a usually inadequate quantity of water to the growing town. The Tanner's Work also appears (as 'Tan Yard') on Kincaid's map [1784], but was removed during the construction of the Mound a few years later. Far from having anything to do with William Wallace the medieval hero, or even the seventeenth-century master mason of the same name, the 'Walace Tower' shown at far left was originally a *well-house*, as its proximity to the well might tend to suggest; the same corruption of 'well-house' is also known to have occurred in Aberdeen.

The City of Edinburgh was clearly hoping to prove, through this map, its ownership of the entire area marked in green: including 5 acres and 2 roods (5.5 acres) on the north of the drain, and 1 acre, 3 roods on the south of it. The confirmed owners on the south side running up to the High Street owned 12 acres of land, and so of the total area of 17 Scots acres, around 25 per cent was disputed. In addition to the Governor of the castle, there were a number of smaller proprietors – including Dun, Brown, Wilson and the tanners 'Bark Hill' – whose claims to houses, grounds and meadow areas between the drain and Canal Street were all deemed 'Doubtfull' by the Council.

John Lesslie was an experienced land surveyor, who lived in Liberton from 1755 to 1778, and who had worked with the Commission for Forfeited [Jacobite] Estates for nearly 20 years (1755–74). This map, drawn at the large scale of 200 feet to an inch or 1:2,400, shows the value of maps in planning drainage, enclosure and property ownership in this disputed area that, as dry land, had become newly valuable. Although drafted for the Town Council, the plan survives in the archives of W. & A.K. Johnston in the National Library of Scotland,

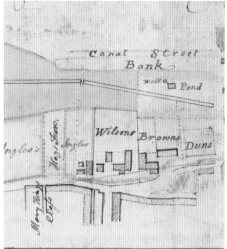

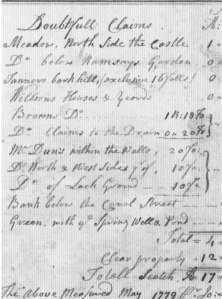

ABOVE. The smaller proprietors including Dun, Brown, and Wilson just west of North Bridge, whose claims were all deemed 'Doubtfull' by the Council.

almost certainly through William Johnston, who became Lord Provost in 1848.

The Scots acre of 1.3 English acres, Scots mile of 1.12 English miles, and Scots pint of 2.98 English pints were all

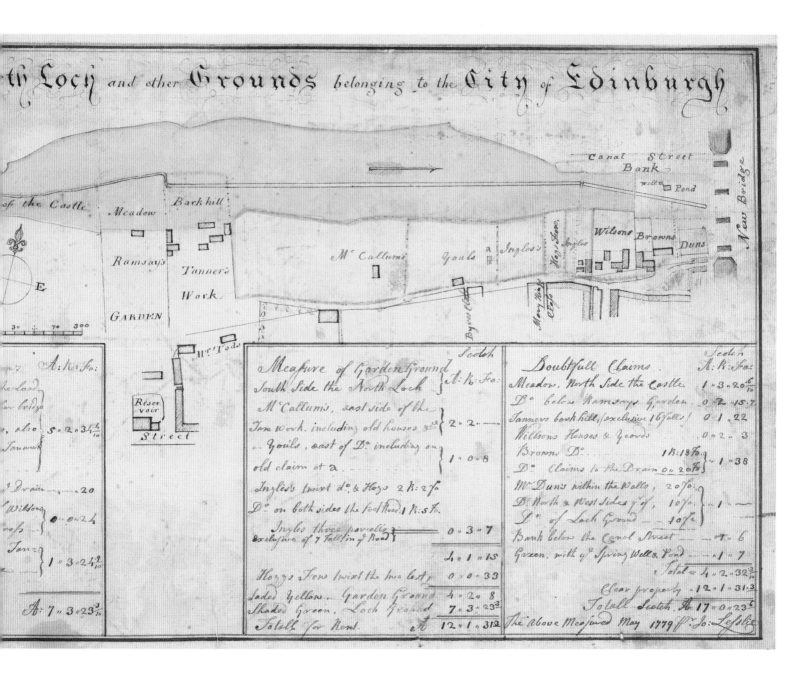

replaced with the familiar Imperial system in 1824, as were numerous other measurements including the ubiquitous ell – a Scots yard of 37 or, sometimes, 42 inches. The mixture of national systems implied in the present plan by the use of English yards and Scots acres, far from being unusual, was probably the norm in Georgian Scotland, where the 'Amsterdam' pound of 1.09 pounds avoirdupois was perhaps as commonly used as any other.

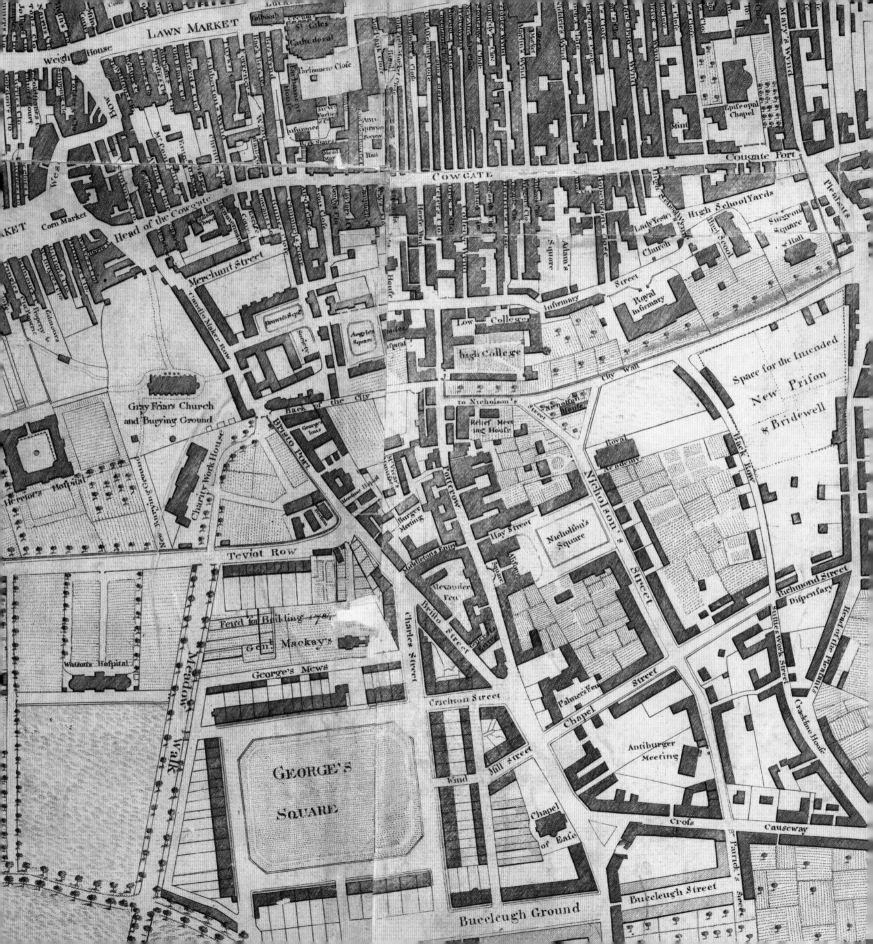

1784

The suburbs on the South Side under construction

By the later eighteenth century, there was a small but growing demand for maps to be included in history books about Edinburgh. The emerging market for these included school pupils, university students, the growing professional classes, and members of learned societies and clubs such as the Society of Antiquaries of Scotland, founded in the city in 1780. This phenomenon encouraged works such as Maitland's *History of Edinburgh* – which had brought Edgar's map [1742] to public attention; Hugo Arnott's *History of Edinburgh* (1779); and three works of the 1780s and 1790s by Alexander Kincaid (fl. 1777–1816).

Of course, more than publishing was affected by the late-eighteenth-century emergence of a middle class so large that it could be treated as a mass market. For instance, it was in this period – and perhaps especially in Edinburgh – that speculative building first became commonplace in Scotland. In earlier eras, no-one would have dreamt of building houses, streets and whole neighbourhoods in the hope that well-to-do residents for them, personally unknown to the builders, could somehow soon be found. A.J. Youngson, in *The Making of Classical Edinburgh*, estimates that only 60 per cent of the houses in Charlotte Square were built 'by people who intended to live in them', implying that 40 per cent of the square was built speculatively; and further, that in less fashionable (but still expensive) new districts of the eighteenth-century city, the percentage of speculators was even higher. In the nineteenth century, speculative building would extend further down the social scale, increasingly targeting the prosperous lower middle class and skilled industrial workers, rather than just the wealthy, as previously.

Kincaid's map shows active development of the New Town and St James Square, as well as the transformation of the South Side. Looking from left to right, we see George Watson's Hospital (the term still signified a private boarding

---

Alexander Kincaid, *A Plan of the City and Suburbs of Edinburgh* (1784)

school with charitable overtones); this had in fact been opened in 1741, but was off the edge of Edgar's maps. In 1766 the builder James Brown, who had previously developed Brown's Square [1765], began the development of George's Square – named not for a saint or king but for his brother, George Brown of Elliston – on parkland to the south of Ross House. It appears here as mostly finished. Brown's development included Windmill Street, which from 1598 had been the site of a windmill for pumping water from the Burgh Loch on the Meadows to a brewery in the vicinity of Teviot Row, via a 4.5-inch diameter lead pipe.

Buccleuch Ground (now Buccleuch Place) was also laid out by Brown, with building just beginning by 1784. Although the land was not owned by the Duke of Buccleuch, the naming was perhaps intended as a compliment, as both the duke and George Brown were members of the Poker Club from 1776. Though many clubs and associations of this type had their origins in the new coffee-house culture of the Restoration era, the eighteenth century was their heyday as a nexus of politics, writing, the arts and a wide range of projects, both practical and impractical. The 66 members of the Poker Club in 1768 included the philosopher–economists Adam Smith and David Hume; the moderator of the Church of Scotland, 'Jupiter' Carlyle; and Henry Dundas, a Tory politician so powerful that he was sometimes referred to as 'Henry the Ninth', as well as two colonels, a surgeon and 21 lawyers. The modern or 'social' type of Freemasonry, as distinct from lodges actually governing the building trades, also came to prominence in this period.

The Antiburgher Meeting House near the top of Nicolson Street provided a place of worship for persons who opposed the Burgher Oath, introduced in 1747, which required holders of public offices to confirm their approval of the Established Church. In the nineteenth century, this became the United Presbyterian Church and today it is the Southside Community Education Centre.

Further east, the 'Space for the Intended New Prison or Bridewell' reflects the Town Council's need to replace the Old Tolbooth by St Giles, which was a truly dreadful place of incarceration, finally pulled down in 1817. Bridewells were so named after the House of Correction by the Holy Well of St Bride, off Fleet Street in London, which had functioned as a prison from the sixteenth century. In fact, the Edinburgh Bridewell would not be constructed until 1796, and on the west side of Calton Hill. In its place, Richmond Street and what would later become Adam Street were laid out in this area by 1800.

Beyond Bell's Brewery to the east lies St John's Hill, where James Hutton (1726–1797), the famous geologist, physician and naturalist, lived at this time. A year after this map was printed, Hutton read his seminal paper on the Earth's geology to the Royal Society of Edinburgh. This was published in 1788 as the *Theory of the Earth; or an Investigation of the Laws Observable in the Composition, Dissolution, and Restoration of Land upon the Globe*. Transposed to the field of biology, Hutton's ideas were a major influence on Darwin's theory of evolution.

Kincaid had become His Majesty's Printer in Scotland in 1777, a title inherited from his father, also Alexander Kincaid, who had printing offices at the foot of Old Fishmarket Close, and who had been Lord Provost of Edinburgh in 1776–77. The younger Kincaid published *The History of Edinburgh, from the Earliest Accounts to the Present Time; by Way of Guide to the City and Suburbs* in 1787, which included an updated map engraved by Hector Gavin showing amongst other things the new South Bridge. Kincaid also published *A New Geographical Historical and Commercial Grammar* (1792) and *The Travellers Companion through the City of Edinburgh and suburbs* (1794). He lived on North Richmond Street after it was laid out in the 1790s, while his printers' ink manufactory was in Fountainbridge. Kincaid had his map engraved by Edinburgh-based John Beugo (1759–1841); this is one of his earliest known works. Beugo went on to engrave maps for William Gordon's *A New Geographical Grammar, and Complete Gazetteer* (1789), as well as the famous portrait of Robbie Burns which appeared in the Edinburgh edition of *Poems, chiefly in the Scottish dialect* (1787).

1790

'La Nature à Coup d'Oeil': Edinburgh's first panorama

Robert Barker (1739–1806) was born in County Meath, Ireland, but was resident in Edinburgh's High Street from 1786, earning his income as a portrait painter and teacher of perspective drawing. He is credited as the inventor of the panorama, patenting in 1787 'an entire new Contrivance or apparatus called "La Nature à Coup d'Oeil" for displaying Views of Nature at large'. This was a circular building, accessed from below and naturally lit from the top, carefully confining the visitors' perspective to the painted image alone.

During his years in Edinburgh, Barker quickly realised the advantages of Calton Hill for his 360-degree perspective, and used the Guard Room in Holyrood Palace as his studio. The 25-foot-diameter painted panorama was a runaway success when exhibited in Edinburgh at the Archers' Hall (headquarters from 1777 of the king's Scottish bodyguard of bowmen, established in the time of James VII) and then the Assembly Rooms, and went on to popular acclaim in Glasgow and London. This image detail shows one of the central panels from a reproduction of the painting, aquatinted on six sheets by J. Wells and published in 1790.

In the far left background of the detail, Heriot's can be seen just behind St Giles, with the Adam brothers' Royal Exchange buildings slightly nearer to the viewer. In the foreground, in front of the original North Bridge in the Old Calton Graveyard, is the circular tower of the David Hume Mausoleum by Robert Adam, constructed in 1777; and to its right, the pretty spire of the 1743 Orphans' Hospital, demolished in 1845 to make way for the railway. With this hospital also went the clock of the Old Town's Netherbow Port, which had been put there for safekeeping on the latter's demolition in 1764.

In the middle distance to the right of a vertical crease in the paper is the West Church/St Cuthbert's Church, rebuilt with a new steeple in 1789. Beneath the castle we can see the

---

Robert Barker, Panorama of Edinburgh from the Calton Hill (c.1790)

early state of the Earthen Mound – as the as-yet unpaved Mound was still known, and indeed still called on maps as late as the 1880s. An estimated 1.5 million cartloads of earth from the levelling of the New Town site were used to construct the Mound over a period of half a century beginning in 1781. Other aspects of the making of the New Town are also shown, with George Street – marked by the exaggerated octagonal spire of St Andrew's Church – only about two-thirds finished, and no further new buildings to the north of it at this time, though Princes Street is essentially complete. The fat chimney-stacks and fairly shallow roof-peaks of the mid Georgian period, while correct in the foreground, are perhaps the product of wishful thinking in the case of the Old Town: where steep-roofed, narrow-chimneyed buildings of the Stuart era are common enough even today.

In the central foreground at the north end of North Bridge are the high buildings at the back of Shakespeare Square. The domed building to the left of St James Square is Robert Adam's General Register House, looking unexpectedly like a Spanish church following its initial phase of construction in the mid 1770s; unusually, it would remain incomplete until a second phase of construction 50 years later, but this is perhaps less surprising when we consider that its total cost was £80,000 (up to £275 million in modern terms).

Barker also painted a panorama of Edinburgh from St Giles's steeple, and a (larger) panorama of London in 1790–91. From 1792 he was able to construct his special Panorama Building in Leicester Square. As built, it included two panorama paintings simultaneously: one 90 feet in diameter in a 'large circle' at ground level, and the other, 50 feet in diameter in an 'upper circle'. Crowds flocked to it. Barker would go on to paint several more panoramic views of British towns including Bath, Brighton and Windsor, as well as various naval triumphs. His success paved the way for other, technically based 'prodigy' painters and panorama operators in the early nineteenth century, including John Banvard of New York and the Marshall brothers of Glasgow. Moving panoramas or 'myrioramas', installed on vertical rollers in a proscenium and accompanied by live narration and (sometimes) sound and lighting effects, became one of the most lucrative forms of mass entertainment in the Victorian period, and an important antecedent of cinema. They remained popular in Edinburgh perhaps longer than anywhere else: with the Poole family showing them in the Synod Hall, Castle Terrace, regularly from 1906 until 1928 – two years *after* the city's first 'talkie' was exhibited in the same venue.

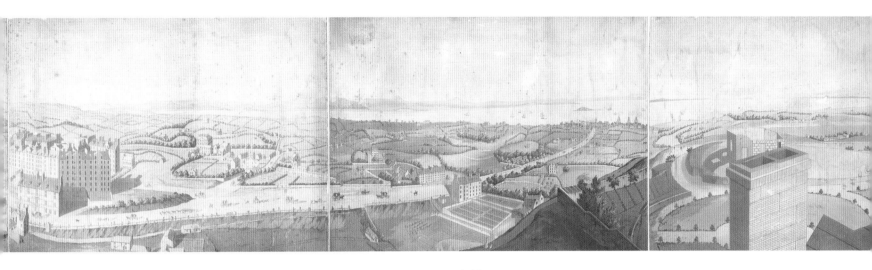

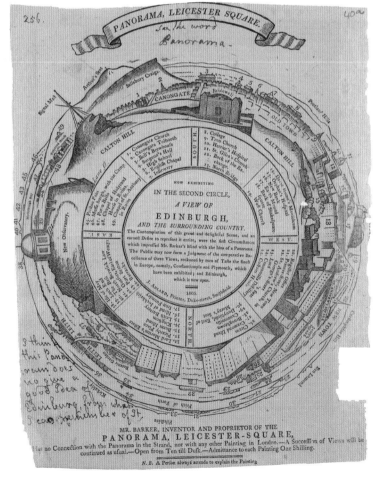

ABOVE. Barker's original 360-degree panorama was 25 feet in diameter. This aquatint reduction measures 16 x 130 inches.

LEFT. An 1803 advertisement for the Edinburgh panorama's exhibition 'in the second circle' in Leicester Square, highlighting its essential features with a numbered key.

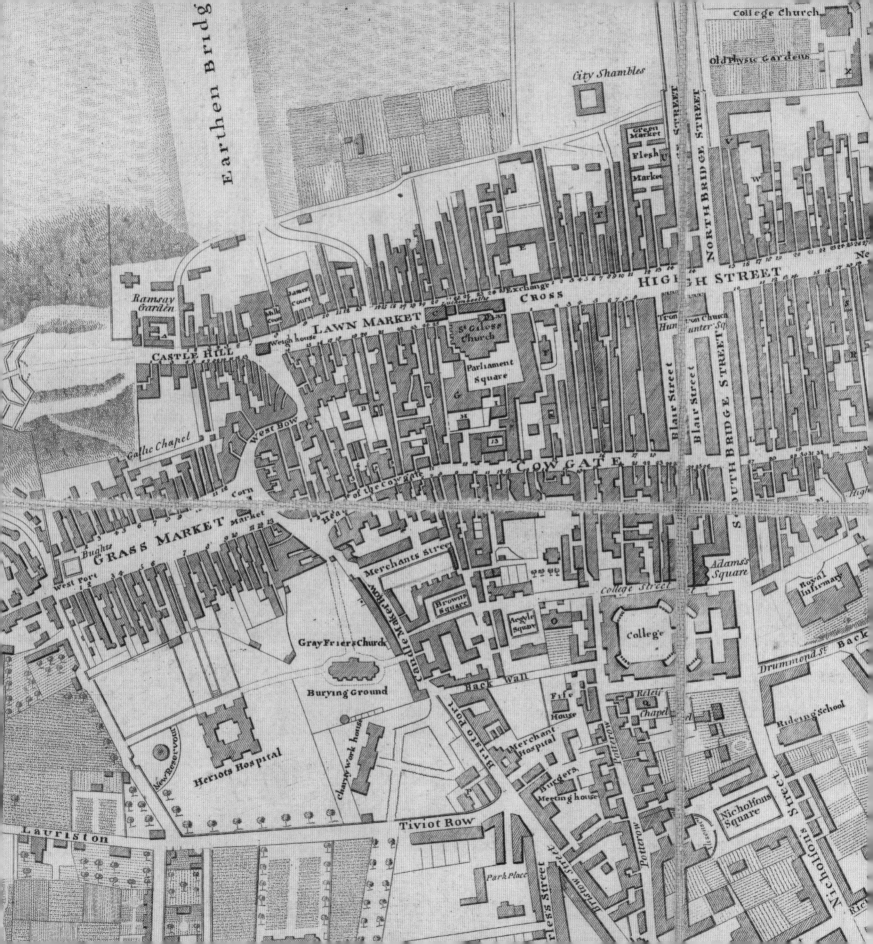

1793

Old College and the new South Bridge

Dissected, mounted on linen and folded in a slip case in a convenient form for the traveller, this carefully executed plan by Thomas Brown and James Watson numbers all the closes and wynds along the High Street, Cowgate and Grassmarket. Although it anticipates certain developments such as Old College and the Bridewell on Calton Hill that were not yet constructed, it is a useful record of recent building activity.

The largest construction project of the previous decade was South Bridge: 1,075 feet long and carried on 22 arches, all apart from the central one now concealed by the substructure of the buildings. Although alternative proposals had been put forward to bridge the Cowgate, the completion of North Bridge in 1772 provided the obvious line to take to the south, and the scheme was vigorously promoted by James Hunter Blair after he became Lord Provost in 1784. Following the necessary Act of Parliament, the foundations of South Bridge were laid by August 1785, and the work was completed in 1788. Along with its adjacent buildings, the bridge involved the demolition of several wynds, including Kennedy's Close, Peebles Wynd, Marlin's Wynd and Niddry's Wynd off the High Street, and Sawers Close in the Cowgate. Robert Adam, Robert Kay, Alexander Laing, James Brown and John Baxter all played roles in the design of the South Bridge scheme, but the exact nature of each man's contribution seems likely to remain permanently shrouded in obscurity.

A number of grand buildings were also demolished in the process, including the Black Turnpike on the High Street – 'one of the most sumptuous edifices of the old town' – and the civic palace of Nicol Uddert or Edward, where James VI had lodged in former times on the west side of Niddry's Wynd. The Tron Kirk was truncated on three sides: on its east side by the bridge itself, on its west for the new Blair Street, and on its south for the new Hunter Square, both named after the Lord Provost, who died in 1787 before the full scheme was

---

Thomas Brown and James Watson, *This Plan of the City Including all the Latest Improvements* (1793)

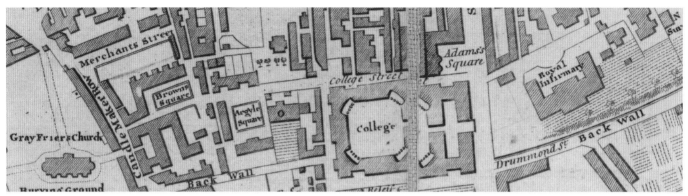
The South Side from Greyfriars to Edinburgh University, including Robert Adam's original two-courtyard plan for Old College.

complete. The shops that fronted the new South Bridge were initially built one over the other, above and below street level, with steps from the bridge leading both downwards and upwards.

Further to the south we can see Robert Adam's original planned layout for Edinburgh University's Old College, at this time only partly built on its eastern side. Adam had put forward alternative and more grandiose proposals for both South Bridge and the College which were not adopted, but he successfully sued the city for costs, and in the end was awarded the College architectural work, albeit with a new brief. As shown here, the plan included two courtyards, an eastern oblong for the professors, and a western square for the library, museum, graduation hall and teaching rooms, with a chapel in between. Adam died in 1792, the year before this map was published, and further work was delayed until 1815 due to the financial stresses of the Napoleonic Wars (which included the innovation of an income tax of 10 per cent on incomes over £200 in 1799, among other taxes). Whilst maps in these years usually show a complete university building, in practice it was only occupied on its eastern side, and the remaining parts 'stood in a condition of hopelessness as to the probability of its ever becoming winged with the other elevations of the original plan'. Fortunately an Act of Parliament in 1815 granted £10,000 per annum to the completion of the building, and under revised designs by William Henry Playfair, the building was finished in stages to 1827, with its dome added in 1879.

Just to the west of Heriot's Hospital, we can also see the 'New Reservoir'. With the construction of the New Town, the original water supply for the city – from Comiston by lead and wooden pipes to a small reservoir in Castlehill – became insufficient. However, an Act of Parliament back in 1758 had stated that the city was not supplied 'with a sufficient quantity of good and wholesome water'. The bore of the pipes was increased, and the new reservoir constructed at Heriot's in the 1780s fed pipes which led along North Bridge to a secondary reservoir on Multrie's Hill, which in turn fed pipes running along the streets of the New Town. That said, those residents privileged enough to have a domestic water supply at all still had to carry water upstairs from cisterns in the basement of their houses.

Thomas Brown and James Watson were both independent booksellers and stationers and members of the Edinburgh Booksellers' Society, who were based initially in Parliament Close in the 1780s, but who moved to the newly constructed bridges in the 1790s: Brown to North Bridge, and Watson to South Bridge. The engraver of this map, James Archer, was still based in Parliament Close at this time. Revised and updated in 1809 and 1820, the map would gain further popularity through its use by John Wood in his collaboration with Brown [1823].

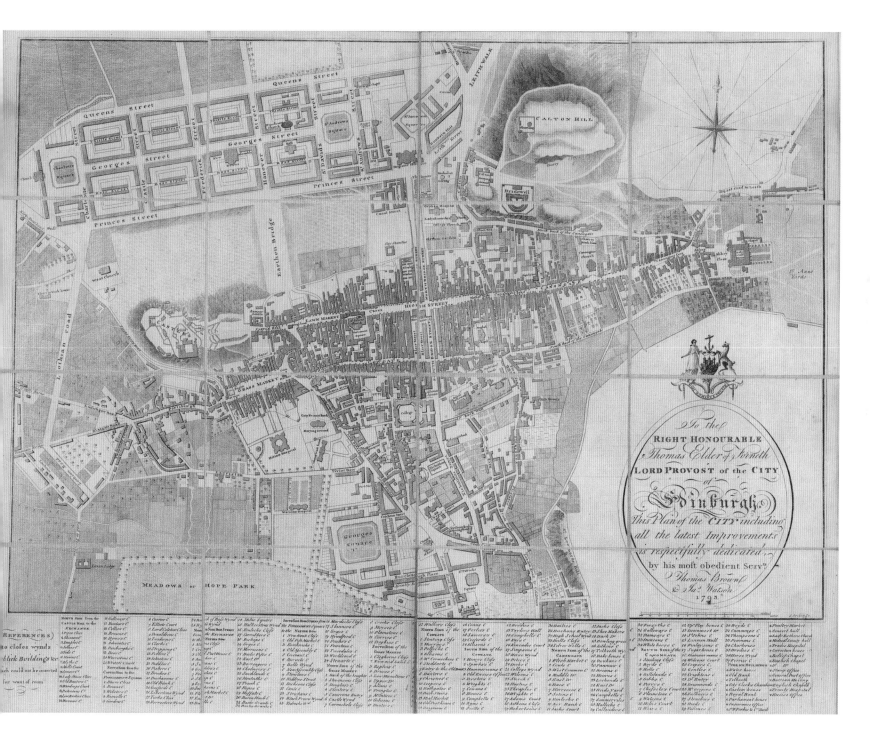

1804

Expanding to the north?

This clear and attractive map is special for a number of reasons, not least that it was the first to include both Edinburgh and Leith on one sheet – a foretaste of things to come, even if municipal amalgamation would not occur until 1920. John Ainslie (1745–1828) was one of Scotland's foremost cartographers, as well as an Edinburgh resident from 1778 until his death, and he lavished particular attention to detail on this map. It is also of inestimable value today as a cadastral plan showing land ownership at the beginning of the nineteenth century, when the Scottish gentry, suddenly barred from the Continent by the Napoleonic Wars, settled in Edinburgh in large numbers, apparently considering it the best domestic alternative to the Grand Tour. The first New Town had more than doubled the city's high-end housing stock in a single generation, but demand still outstripped supply.

This detail focuses on the northern New Town, including Canonmills and Broughton, which were actively developed during the first quarter of the nineteenth century. However, disentangling real from planned streets and buildings is far from easy. Whilst the New Town south of Queen Street was largely complete by this time, and the streets and buildings to the north of it were planned from 1802, many of the latter would not be constructed for another two to three decades – and not always according to the scheme shown here, which Ainslie freely admitted to simply copying from a plan owned by James Jackson, Esq. The unbroken Royal Circus on the left was, in fact, breached by an important diagonal routeway, which prevailed as the natural, historic route to Stockbridge. The mill lades running into Canonmills Loch can be clearly seen, as well as the larger haugh (compare [1766]). James Eyre – whose name lives on in several streets in this area – was a brewer who owned Canonmills House and the neighbouring lands, as we can see here; these were planned for development from the 1820s.

John Ainslie, *Old and New Town of Edinburgh and Leith with the Proposed Docks* (1804)

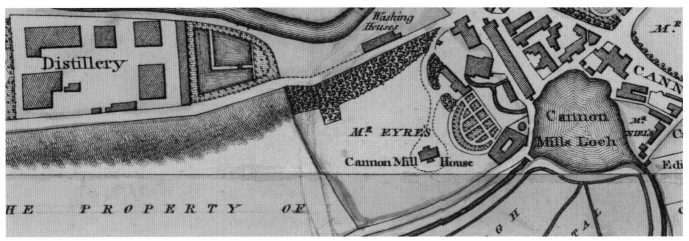

ABOVE. Canonmills Distillery, owned by John Stein, and Canonmills House and Brewery, owned by James Eyre.

OPPOSITE. The expanding New Town engulfed several former villages. Here the new, straight and grand 'Caledonia', 'Anglia' and 'Hibernia' Streets (which would become Scotland, London and Dublin Streets) contrast with the asymmetrical and narrow streets of Old Broughton.

The massive distillery at upper left was a truly industrial concern, founded by James Haig in 1779 and capable of producing an incredible 35,673 gallons of spirits in each 12-hour shift. Despite being Scotland's largest distillery, Canonmills was closed by misguided export legislation in 1788. It was purchased two years later by Haig's principal rival in the Lowland cheap-whisky market, John Stein. During food shortages in 1794, it was attacked by a mob and had to be defended by troops; the mob leaders were caught and transported to Australia. Stein's ran it for another 31 years before selling it back to Haig's in 1825; it ceased to be a distillery after 1840 and was demolished in the 1970s.

Off Drummond Place, 'Caledonia', 'Anglia' and 'Hibernia' Streets – in fact Scotland, London and Dublin Streets – were named to commemorate Ireland's incorporation into the United Kingdom. It is now often forgotten that the names of most streets and other places called 'Union' in Scotland, and of the once-powerful Unionist and Liberal Unionist political parties, referred to this 1801 union of Britain with Ireland, and only indirectly (if at all) to the union of Scotland with England 94 years earlier.

George Heriot's Hospital was still the largest landowner in the area covered by Ainslie's map, holding 376 acres across eight main sites; this was closely followed by the Earl of Haddington, with 368 acres covering Holyrood Park, whose forebear the 1st Earl had been appointed Hereditary Keeper of the park in 1691. The third largest landowner was James Rocheid with 182 acres, primarily in Inverleith, including land that would be leased to George Lauder in 1823 to form the Botanic Gardens; the Rocheids had owned this land for more than two centuries, even though they were generally resident in Germany. The City of Edinburgh was the fourth largest landowner, with 163 acres, and the fifth largest was Trinity Hospital, who had 118 acres east of Leith Walk extending to Lochend. The majority of these owners were effectively trusts, and this would have a significant impact on the timing and nature of property development and building activities as Edinburgh expanded.

Ainslie's map was a magnificent achievement, as well as a costly undertaking, requiring an original survey over an extensive area. The Town Council Minutes for 13 June 1804 record the receipt of the map as a gift from Ainslie, together with a note: 'I have no prospect of being repaid for my labour and expence but by its sale.' The Council were helpful, without being over-generous: ordering the Treasurer to pay Ainslie £20 if he were to supply them with a further six copies of his map for use in public offices.

Ainslie was born and brought up in Jedburgh. Apprenticed to Thomas Jefferys, Geographer to George III, he learned land surveying and cartography whilst producing several county maps of England during the 1760s. Surveying required patronage and private finance at this time, and Ainslie's early years were dogged by misfortune and difficulties in raising subscriptions for county mapping in Scotland.

During the 1780s and 1790s Ainslie's fortunes turned, and he produced a wide array of significant maps – ranging from the whole of Scotland, its counties and coastlines to estates, canals and roads. He worked with Robert Whitworth surveying lines for the Edinburgh to Glasgow canal (see [1817]), and with John Rennie, civil engineer, surveyed the new harbour at Saltcoats and the line for the Glasgow to Ardrossan canal. Mapmaking by this time was quite specialised, often with separate individuals employed in surveying, engraving and publishing; but Ainslie was an impressive all-rounder, often performing all three of these functions himself. He also wrote important texts on related subjects, including *The Gentleman and Farmer's Pocket Book, Companion and Assistant* (1802), on agricultural improvement, and a *Comprehensive Treatise on Land Surveying* (1812).

Like so many others involved in Edinburgh's publishing and civilian mapmaking trades, Ainslie lived initially in Parliament Square. He moved to the New Town in 1788, where he gave a variety of addresses (in St Andrew Street, South Hanover Street and East Rose Street) before settling in Nicolson Street from 1800 onwards. At the time of his death he had amassed some £8,976 – an incredible achievement for anyone at that time, but the more remarkable given the familiar story of poverty and struggle experienced by other non-military land surveyors of his own and earlier generations.

1813

The Canongate in transition

The eleventh and twelfth centuries saw profound change in Scottish society, with the establishment of Continental-style monasteries, the advent of Anglo-Norman feudalism complete with castles and sheriffs, and the founding of the first burghs (including Edinburgh). As part of the foundation of Holyrood in 1128, King David I granted the abbey a burgh with a wide jurisdiction or 'regality', which included the lands of Pleasance, the barony of Broughton and part of Leith. In the early period the boundaries of this burgh were laid down by written charters, but by the early nineteenth century it was both necessary and possible to confirm these in cartographic form: hence this map of the Regality of Canongate with the boundary clearly marked in red.

This attractive and useful map captures Canongate in transition between its former glories as a centre for King James VI's court and the palace society of the late sixteenth century, and its late-nineteenth-century incarnation as one of the worst slums in Great Britain. The great and the good who had frequented Canongate up until the late eighteenth century were departing, and the new Regent's Road that would open in 1817 allowed new access to the city from the west, further bypassing the burgh.

Proceeding clockwise around the map, we see at upper left the Canongate Tolbooth, with its distinctive turrets, which had functioned as a civic centre and prison for the burgh since the late fifteenth century. Already in decline, it would be sold by 1840. The adjacent Canongate Church was opened in 1691, after the former Abbey Church-turned-parish church was re-dedicated to Roman Catholic worship by King James VII. Begun in 1688, the last full year in which the established Church of Scotland was Episcopalian in character, Canongate Kirk is very Scottish (or at any rate very idiosyncratic), with a

---

William Bell, *Plan of the Regality of Canongate Comprising the Liberties of Pleasance, North Leith, Coal-hill and Citidal Thereof* (1813)

massive armorial panel set in an immense curvilinear gable surmounted by a gilded stag's head. The Presbyterian Church of Scotland churches built in the following century were, by contrast, very English-looking, and in many cases little more than copies of London churches built out of local materials; the most obvious example is St Andrew's, George Street (1781) [1790], which was adapted by an officer of the Royal Engineers from the original, unexecuted design for St Martin-in-the-Fields. In a sense, Scotland's civil conflicts of the eighteenth century would be powered by this curious double opposition between the nation's two leading Protestant Churches: with the Establishment being broadly speaking Anglophile but anti-Anglican, and the Dissenters, Anglican and nationalist.

By the early nineteenth century, Canongate kirkyard contained the graves of a diverse array of famous residents including Drs John and James Gregory, Provost George Drummond and Robert Fergusson the poet, as well as a pardoned Jacobite officer who became a count in Sweden; an Episcopalian bishop/historian; a German-born composer of Methodist hymns; and Ebenezer Scroggie, whose gravestone helped inspire Charles Dickens to create the character Scrooge. The Charity Workhouse at the corner of Old Tolbooth Wynd and Calton Road was built in 1761, part of a growing movement both north and south of the border. Under Scottish legislation of the late sixteenth century, 'strong and idle' beggars were to be punished first by branding on the ear, or by death in the case of repeated offences. Workhouses, seen as a more humane alternative, were authorised in 1672 and were operating in the Edinburgh area by 1720.

A little further to the east by Panmure Gardens is Panmure House: a much-modified T-plan tower-house built for the earls of Panmure, where Adam Smith lived from 1774 until his death in 1790. He too is buried in Canongate kirkyard. Whitefoord House, the cross-shaped building almost opposite Queensberry House, had been built for Sir John Whitefoord in 1769 on the former site of Seton's Lodging, and by the time of this map had passed into the hands of Sir William Macleod Bannatyne, raised to the bench as Lord Bannatyne in 1799. Following his death in 1833 it became a type factory.

Holyrood House by this time was rarely visited by royalty, and the abbey grounds still functioned as a sanctuary for debtors [c.1690].

On the south side of the Canongate by Horse Wynd was Lothian Hut, an elegant town house constructed in 1750 by William, 1st Marquis of Lothian; at the time of this map, it was the residence of the philosopher Dugald Stewart, who would eventually join Adam Smith and the others in Canongate kirkyard. In 1825, the immediate area of the Hut was acquired by William Younger II, and was converted into maltings before being demolished to make way for the Abbey Brewery. The nearby St Ann's Brewery north of Holyrood, and Young Brewery to the west, hint at the shape of things to come [1946].

Queensberry House, which survives today (albeit largely surrounded by the new Scottish Parliament), was built for Charles Maitland in 1679–81, and passed to the Dukes of Queensberry by 1686, who maintained it as a magnificent and sumptuous residence. It continued to house persons of high rank even after its eighteenth-century conversion into flats; but in 1808, amid the creeping industrialisation of the entire area, it was sold to the Board of Ordnance and converted into the infantry barracks we see here, with a parade ground behind.

Milton House to the south of the High Street had been built in the first half of the eighteenth century in the garden of the Duke of Roxburgh by Andrew Fletcher of Milton; his uncle Fletcher of Saltoun owned all the land below the South Back of the Canongate/Holyrood Road (as shown). By this time in the early nineteenth century it had become that categorically new thing of George III's reign: a legally tolerated Roman Catholic school.

Beneath and alongside these grand houses, we can trace the 'Common Sewer from Cowgate' joining the larger 'Common Sewer from the Pleasance' and then continuing east to terminate in the 'Reservoir for Mud and Manure', just south of Holyrood Abbey. The attractively laid-out palace gardens and royal bowling green were presumably well fertilised – but at what cost to the noses of those living nearby?

By 1856, the Canongate jurisdiction was finally merged with Edinburgh via the Municipal Extension Act.

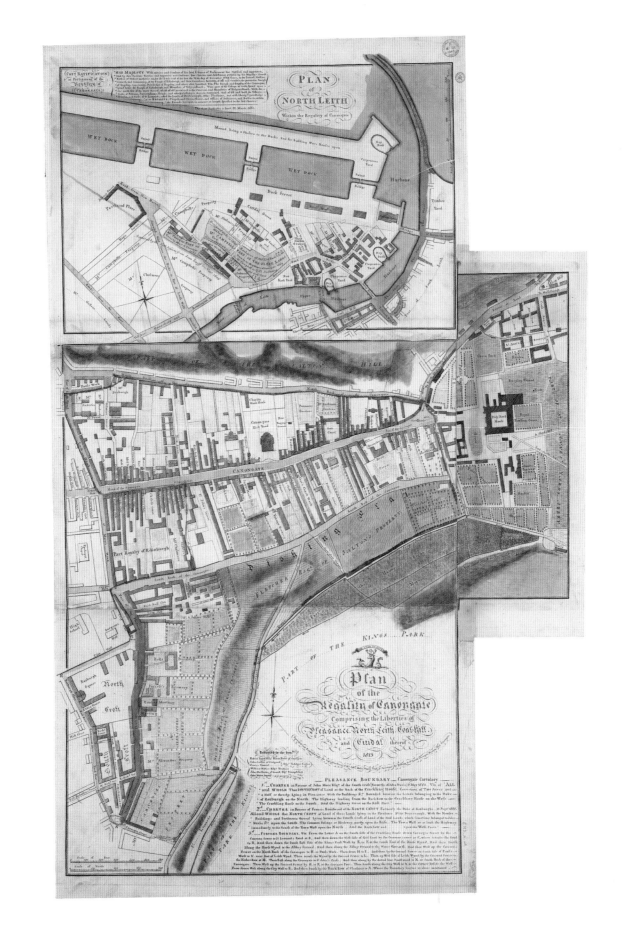

1815

The Forth in the Napoleonic era

This fascinating Admiralty chart of the Forth essentially codifies a system of navigation based on the sizes and relative positions of shore-based landmarks as seen from certain points offshore. 'To Sail over the flat of the Drum Sand into the South Channel', for instance, a shipboard observer should 'keep Dalgety Parsonage House twice its apparent breadth to the Westward of the Hay Stack Rock'. Known to coastal skippers and fishermen for centuries, this method was virtually foolproof on a clear day, but useless at night or in fog. Though not nearly as selective as it might have been, the chart definitely favours tall, prominent features ranging from massive hills like Arthur's Seat and North Berwick Law to steeples, towers and castles, as well as smaller structures that are unlikely to be mistaken for others, including lime-kilns and dovecotes, making it a useful 'prequel' to the Ordnance Survey maps of the mid century. Along with depth-readings and the local variation between magnetic north and true north, it provides a fairly thorough record of structures that were purpose-built to aid navigation, communications and defence: the 'Signal House' on St Abb's Head and the 'Naval Signal House' a little farther west; the 'black buoy', 'red buoy' and 'mast buoy' of the South Channel; Inchcolm's artillery battery; the Inchkeith lighthouse, completed in 1804; and Leith's signal tower and martello tower.

The purpose of martellos, built in many parts of Britain and its empire beginning in 1795, was to house the beacons that would be lit to warn of seaborne attacks, along with a single artillery piece and approximately 20 soldiers. Their curious name was the product of an error on the part of Glasgow-born General Sir John Moore (1761–1809) who, during a British raid on Corsica in 1794, had to assault a sixteenth-century signal tower from the landward side because two Royal Navy warships were unable to blast it into submission from a range of just 150 yards. This tower directly inspired Britain's anti-invasion beacon chain. Upon being

George Thomas, *Survey of the Frith of Forth* (1815)

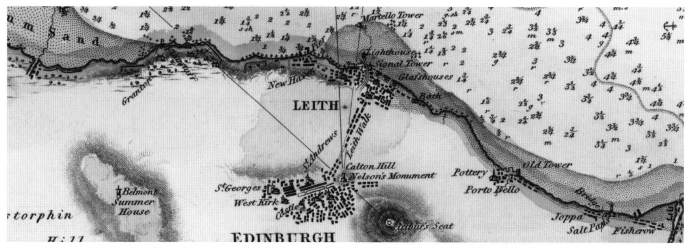

These three transits from the Nelson Monument, intersecting with Arthur's Seat, North Leith Church, and Leith's martello tower, allowed mariners in the Firth of Forth to avoid the Drum Sands of Cramond and the Gunnet Rocks near Inchkeith.

informed that the tower was called *Mortella* ('Myrtle'), Moore scribbled the name down with the first and last vowels transposed, little suspecting that the result (meaning 'Hammer') would become a permanent feature of the architectural history of his country, as well as of the British colonies of Nova Scotia, Quebec, New Brunswick, Bermuda, Barbuda, the Virgin Islands, Jamaica, Trinidad, Sierra Leone, South Africa, Ceylon (Sri Lanka), and even British-occupied parts of the Mediterranean such as Minorca and Sicily. The armed forces of the early US republic, while at war with Britain 1812–15, copied martellos they had seen in Canada and replicated them at sites ranging from Portsmouth, New Hampshire, in northern New England to Tybee Island, Georgia, in the Deep South.

Though more than 150 martellos were built in England, Ireland, Wales and the Channel Islands, Leith's example (completed 1809) is one of just three in Scotland, and the only one on the Scottish mainland. Aside from the fact that the main thrust of any French invasion was expected to strike Kent and Sussex, this relative dearth may have been because Scottish ports from the Solway to Banff had already armed themselves to the teeth against the French and, latterly, Yankee privateer attacks of previous wars. Known locally as the Tally Toor, Leith's martello would return to service as an anti-aircraft emplacement during the Second World War.

Constructed 1807–16 on the former site of a naval signal mast, and itself a perfect feature for ship-to-shore sightings, Calton Hill's six-storey Nelson Monument was a memorial to Vice Admiral Lord Nelson (1758–1805): a further reminder that this chart was completed in the final year of the French Revolutionary and Napoleonic Wars, a two-decade global struggle that had cemented Great Britain's position as 'top nation', albeit at a terrible human cost on the battlefield, at sea and at home. It was only after the commencement of these wars that the Hydrographic Office of the Admiralty was first officially established, under Newhailes-born Alexander Dalrymple (1737–1808), first Hydrographer of the Navy and a chief proponent of the theory that there was a vast undiscovered continent lying in what we now know to be the open water of the South Pacific. Apart from the formal gesture of the establishment of the Hydrographic Office, however, wartime conditions seem to have hampered rather than accelerated chartmaking in Scottish waters, with private enterprise continuing to fill the void throughout the war years and beyond.

Mapmaker George Thomas had something of a 'Robinson Crusoe' experience early in life: while an apprentice on a

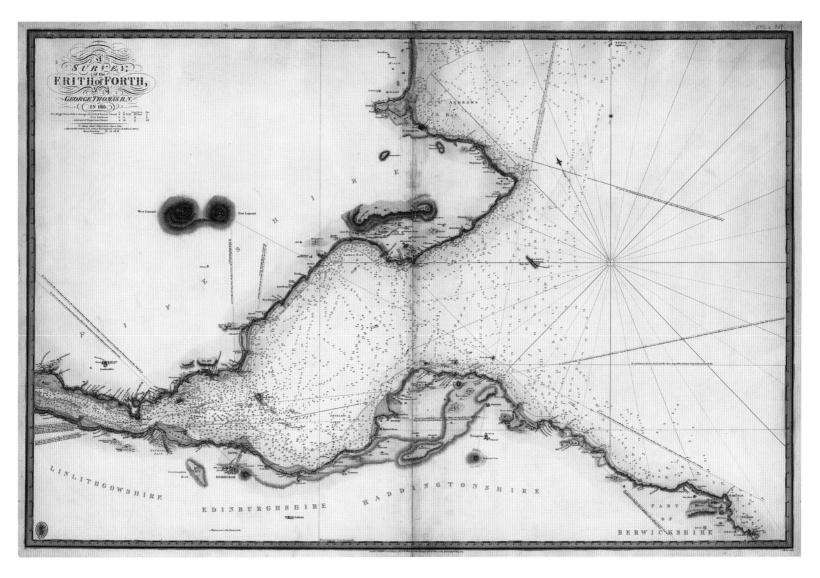

whaler, he was wrecked in the Juan Fernández Islands 400 miles west of Chile. (The Scottish sailor who inspired Defoe's *Robinson Crusoe*, Alexander Selkirk, had been stranded on the neighbouring island from 1704 to 1709.) Thomas spent his time usefully, collecting valuable seal skins. He was ultimately rescued, only to be forcibly enrolled in 1803 in the Royal Navy, which unlike the pre-1916 British Army operated a system of conscription, known as the 'press'. He again made the best of difficult circumstances, so impressing his superiors with his aptitude in navigation and surveying that he was appointed Head Maritime Surveyor for Home Waters in 1810. He would spend the next 36 years almost continually at sea, mostly surveying the Scottish coasts from Berwickshire to Angus (1815–23), then Shetland (1825–35) and Orkney (1837–46). Thomas had a meticulous attention to detail and incomparable stamina, and his surveys of the Northern Isles – the Admiralty's first – were not fully superseded until recent times.

1817

'Containing all the recent and intended improvements...'

Measuring 63 × 57 inches, this finely detailed and beautifully engraved plan by Robert Kirkwood (1774–1818) is one that can be returned to again and again, always with renewed pleasure and yielding fresh discoveries. Though this detail focuses on the Tollcross area, the whole map covers a broad swathe of country some way around the built-up area, from Granton in the north and Morningside in the south across to Seafield, Duddingston and Craigmillar in the east.

Town Council records confirm that on 18 October 1815 Robert Kirkwood applied for access to the plates and plans of John Ainslie [1804], to make a new plan that was up to date. Access was granted, and while this would have been helpful for the central part of Kirkwood's plan, surveying was necessary for all the areas beyond, as well as to achieve the central purpose of updating Ainslie. Kirkwood worked closely with the council, gaining access to unpublished proposals, and generally showed these planned features in red, or described them as 'Proposed'. He also dedicated the map to Lord Provost William Majoribanks, 'under whose able and popular administration, the execution of these magnificent improvements was begun'.

One of the key improvements in question was the Union Canal, newly authorised in 1817 to run 31 miles eastward to Edinburgh from the Forth and Clyde Canal near Falkirk. At this time there were three rival proposals by different engineers, all illustrated here. Robert Stevenson [1819b] had submitted two more northerly routes to Leith, one running along the Water of Leith, and the other taking a course through Princes Street Gardens, similar to that of the future railway, and then turning north around the east side of Calton Hill. Stevenson's most bitterly jealous rival, John Rennie, had proposed a more southerly route through Morningside, terminating at the east end of the Meadows. The more central course proposed by Hugh Baird (with support from Thomas

---

Robert Kirkwood, *Plan of the City of Edinburgh and Its Environs* (1817)

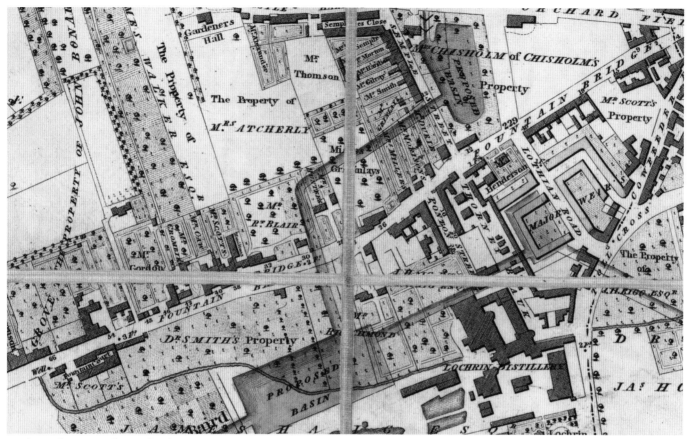

Hugh Baird's proposed route for the Union Canal, implemented by 1822 with its terminus at Port Hopetoun to the east of Semple Street.

Telford) was the one actually approved, and it is shown in blue, terminating at Port Hopetoun. Work started on this in March 1818 and it was complete four years later.

The largest building shown on the map is the Lochrin Distillery, reflecting the accelerating pace of Britain's industrial revolution as much as the new-found importance of Scotland's best-loved export. Lochrin was established in 1780 by John Haig, brother of the James Haig who had founded Canonmills Distillery [1804] the previous year. After similar vicissitudes, Lochrin became an ironworks in 1860.

Kirkwood's map also gives useful cadastral information, much like the Ainslie map of 13 years earlier, but with a very different ranking of top landowners. This is due primarily to the wider area it covers. The largest five landowners were the Earl of Haddington, whose full extent of Holyrood Park was 583 acres, followed by the Marquis of Abercorn's Duddingston Estate with 479 acres. Third was William Henry Miller, who owned 380 acres at Craigentinny and Seafield; fourth, Heriot's, with 376 acres; and fifth, James Rocheid of Inverleith with 353 acres. Altogether, the map names 632 landowners, who collectively owned 6,371 acres, but with a heavily skewed concentration: the top 30 landowners owned 76 per cent of this land. Even so, this was a very large number of very small holdings in comparison to almost any randomly chosen area of equivalent size elsewhere in the country, at a time when 20 freeholders in a parish was considered a great

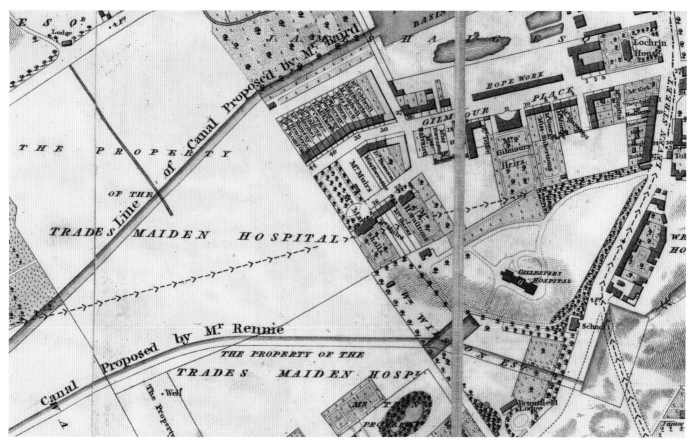

John Rennie's proposed route for the Union Canal was the most southerly of the three, seen here running here across Bruntsfield and on to the Meadows via a series of locks.

many, and estates of over 1,000 acres were commonplace. The right to vote in parliamentary elections at this time was based on ownership of land worth £35, and the average number of qualified voters in a Scottish county was just 71.

Kirkwood was born in Perth, but moved with his father James to Edinburgh in 1785 or 1786. James was licensed originally as a clock- and watchmaker in Perth, but his practical metalworking skills enabled a natural move into engraving; his earliest known work of this type was a plan of Perth in 1774, published in *The Muses Threnodie*.

The 'Great Fire' of 15–17 November 1824, which devastated a wide area between the High Street and Cowgate, destroyed the Kirkwood business premises at 19 Parliament Square. The fire began on the second floor of a printer's workshop at the head of Old Assembly Close, but because of the narrowness of the wynds it was impossible to get fire engines near, and it quickly spread. At times it seemed as if the fire would be brought under control, but new sparks kept spreading it further. The spire of the Tron Kirk was brought down at this time, along with high tenements on the south side of Parliament Square. Thereafter the Kirkwoods sensibly relocated to South St Andrew Street in the New Town, but their stock of copper plates, including the ones for this map, had been lost in the fire. The firm of Alexander Kirkwood & Son still survives today in Albany Street as medallists and engravers.

1819a

A curious double perspective on the east end of Princes Street

This uniquely impressive and beautifully engraved graphic, as its title suggests, manages to flatten all the elevations of buildings in the New Town into a conventional overhead plan, thereby compressing three dimensions into two. Students of early-nineteenth-century Edinburgh are indebted to the Kirkwoods for their high-quality surveying and engraving work, and for the family of maps they published between 1817 and 1819 depicting the city in multiple ways, particularly since the market for these was limited. We know from the Town Council records that Kirkwood appealed for financial aid in publishing this plan and elevation on 29 April 1818, but it was refused.

The original development of this particular area at the east end of Princes Street – which from 1896 was redeveloped as the North British Hotel – caused major litigation culminating in a House of Lords decision of 1818 which fundamentally changed the future nature of feuing and housebuilding in Scotland. In the 1770s, the Town Council had feued land extending westwards from North Bridge for building workshops. But 16 proprietors of the New Town, including David Hume, objected, claiming that this ran counter to James Craig's plan which promised unfettered views across to the Old Town and castle. The compromise reached in 1776 allowed the new buildings to be completed, but decreed that all ground further westward from these to Hanover Street would be 'kept and preserved in perpetuity as a pleasure ground'.

Whilst this quelled the initial dispute and halted any further westward expansion of building on Princes Street Gardens, the decision was in fact overturned by the House of Lords. Although precipitated by building developments elsewhere in the New Town, the 1818 ruling declared that feuing plans alone could not be used to dictate development, and that clear written burdens and restrictions on development

---

Robert and James Kirkwood, *Plan & Elevation of the New Town of Edinburgh* (1819)

had to be specifically stated. From this time onwards, therefore, feu charters involved increasingly detailed written stipulations, which we find for much of the development of the northern New Town, Moray Estate and West End, even though the original plan for the New Town, with its open spaces and parks, had been successfully defended on an interpretation of Craig's plan alone.

The area depicted housed a very different class of inhabitants from the New Town at large. John Home, who acquired the land in the 1770s, was a coachbuilder, and by this time his large coachworks illustrated here were occupied by Messrs Learmonth & Co. Although there were a few hotels nearby, the street that formerly ran between the coachworks and North Bridge, St Anne's Street, was occupied in 1816 'exclusively by keepers of ale-houses and small shops, or by chairmen, porters, and common mechanics; and in particular by a numerous and exalted colony of operative tailors', housed in 'a range of dirty and deformed chimney tops . . . in which the most curious eye could scarcely discover any feature of the sublime or beautiful'. An Act of 1816 widened North Bridge, opening up the view to Register House, and the houses of St Anne's Street were demolished. Kirkwood's plan here shows the new buildings, constructed further back to add to the width of the bridge. In compensation, the people of Canal Street were granted a serpentine road of gentle ascent to Princes Street, clearly shown here, which also allowed access to the public markets from the New Town.

The many buildings depicted are realistic in both form

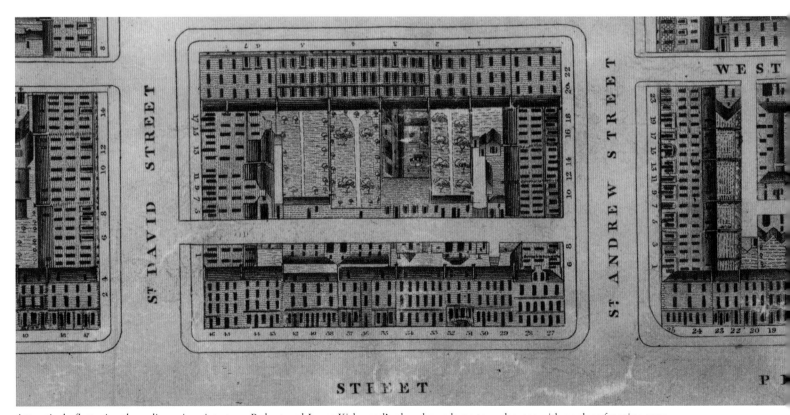

Attractively flattening three dimensions into two, Robert and James Kirkwood's plan shows large town houses with gardens fronting onto Princes Street and St Andrew's Square, and stables and mews houses running between them.

and detail, and show an easy grasp of the distinctions between residential and commercial architecture, as well as between the mid and late Georgian styles. By some unwritten rule of the period, a dwelling generally had a width consisting of an odd number of windows, and a business usually had an even number. The building shown at 27–28 Princes Street – with a width of six windows, two slightly off-centre doors, and a central, street-facing gable containing two windows and a chimney – is typical of Scottish commercial buildings of 1740–80.

At 15 Princes Street was Fortune's Tontine Tavern and Coffee-House, which was acquired through a 'tontine', a peculiar type of lottery (now banned) that originated in seventeenth-century Italy. In 1795 Matthew Fortune, a vintner in Edinburgh, sold shares of £100 apiece (later reduced to £50 and £25); each shareholder would nominate a person 60 years old or older, and the shareholder whose nominee lived the longest would become sole proprietor of the new tavern. The list of shareholders included the dukes of Hamilton and Buccleuch and the earls of Eglinton, Cassilis, Dalhousie, Breadalbane and Hopetoun, as well as Henry Dundas (afterwards Lord Melville) and Robert Dundas, Lord Advocate. By 1820, the number of nominees had fallen to nine and they agreed to divide the property between them.

To the east can be seen the Theatre Royal in Shakespeare Square, constructed in 1769 and demolished in 1860–61 to make way for the new Post Office.

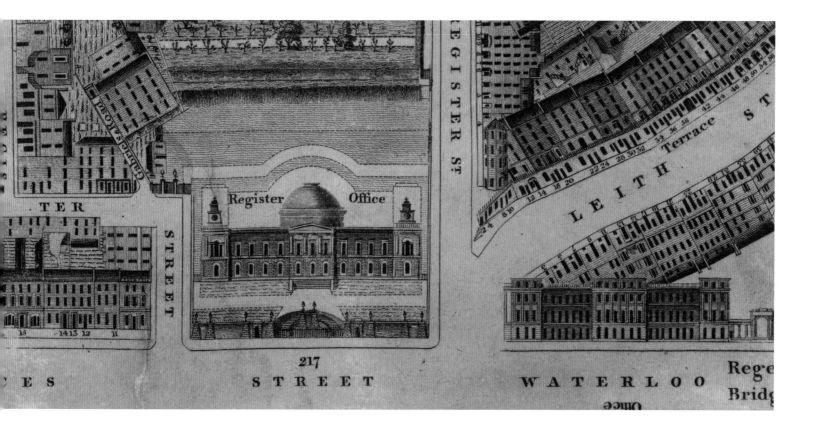

1819b

'There are no stars so lovely as Edinburgh street-lamps'

Robert Stevenson (1772–1850) is remembered chiefly for two things: being the grandfather of world-renowned novelist Robert Louis Stevenson, and designing and building the Bell Rock Lighthouse on the deadly Inchcape Rock, 11 miles off Arbroath, from 1807 to 1811. The Bell Rock light, astonishing achievement though it was, was just one of many faultless Stevenson civil engineering projects throughout Scotland and England into the 1840s, including at least 14 other lighthouses but also canals [1817], harbours [1834b], roads and bridges. Stevenson was directly responsible for the design and construction of Edinburgh's Regent Road, London Road and the dramatic Regent Bridge in Waterloo Place. He was also Scotland's leading expert on horse-drawn railways, and in 1818 proposed a line serving Edinburgh, Leith and the Midlothian coal fields. Though financially unworkable at the time, this scheme eventually succeeded as the Edinburgh and Dalkeith Railway in 1831.

It is poorly remembered, if at all, that Stevenson's marine engineering career was underwritten by the building of Edinburgh's New Town, for which Thomas Smith (d. 1815), who was Stevenson's first employer, stepfather and father-in-law, supplied all the street lighting in 1804. The streetlights of Leith followed in 1810. Stevenson's close relationship with Smith and successful experiments in this seemingly mundane arena qualified him as an expert in the field of artificial lighting as a whole, and contracts to create or improve various lighthouses quickly followed. Baxter's Place, the location of the Smith–Stevenson home and workshops, is clearly shown at upper right on this splendid map with its charmingly old-fashioned spelling ('Armory') and handwriting. Though he was born in Glasgow, Stevenson's deep personal connection to the area depicted may go some way to explaining why this particular 'sketch' was crafted with such obvious care and charm, while his relative lack of formal

---

Robert Stevenson, *Sketch of part of the City of Edinburgh and extended Royalty* (1819)

education explains some of its quirkier points of orthography.

This map represents one of several early attempts to plan new, straight and wide access routes into Edinburgh from the west (the eastern approach via Regent Bridge and Waterloo Place having just been completed). The main aims were to bypass both the narrow West Port into the Grassmarket as well as the West Bow's awkward zigzag up to the Lawnmarket. Stevenson's scheme proposed a new straight road in from the Haymarket; a new curved road bypassing West Port; and a diagonal along the lines of Victoria Street, sloping down to the Grassmarket from near St Giles, but without any equivalent of the future George IV Bridge. In the event, Thomas Hamilton's scheme was approved by an Act of Parliament in 1827, but more than a decade passed before it was implemented, while conflicts over land ownership hindered the development of the area west of Lothian Road for a generation. Further east, what would become Johnston Terrace north of the Grassmarket opened in 1836 as the 'New Western Approach', and Victoria Street opened in 1837. West Port was widened only in the 1880s.

Part of the complications in planning these new streets related not just to multiple landowners, but to the area's patchwork of political jurisdictions, including Portsburgh – a free burgh of barony erected in 1649 that included West Port, Tollcross and Bristo. The original boundaries of the City of Edinburgh, its 'Ancient Royalty', had been confirmed in 1785, and focused very much on the Old Town. But with the development of the New Town and West End, adjacent lands (the 'Extended Royalty' referred to in this map's title) were successively annexed. Even so, by 1819, they still occupied only a small inner core of the area that the city would occupy a century later. Portsburgh was not formally annexed until 1856.

Far from taking an inordinate pride in his many achievements, the somewhat puritanical Stevenson was haunted throughout his life by his lack of education, 'polish' and writing ability. He was also a martinet who moulded the Scottish Lighthouse Service on needlessly (indeed almost comically) militaristic lines. A careful reading of his life and

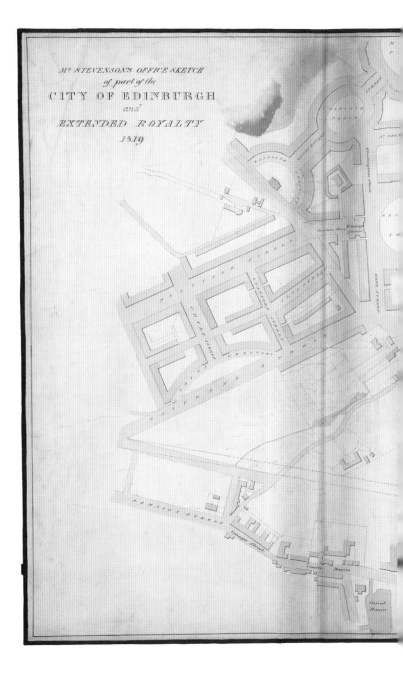

career suggests that he would have thrived as a military officer, in terms of the work needing to be performed, and perhaps secretly wished to have been one, but that he was unsuited to

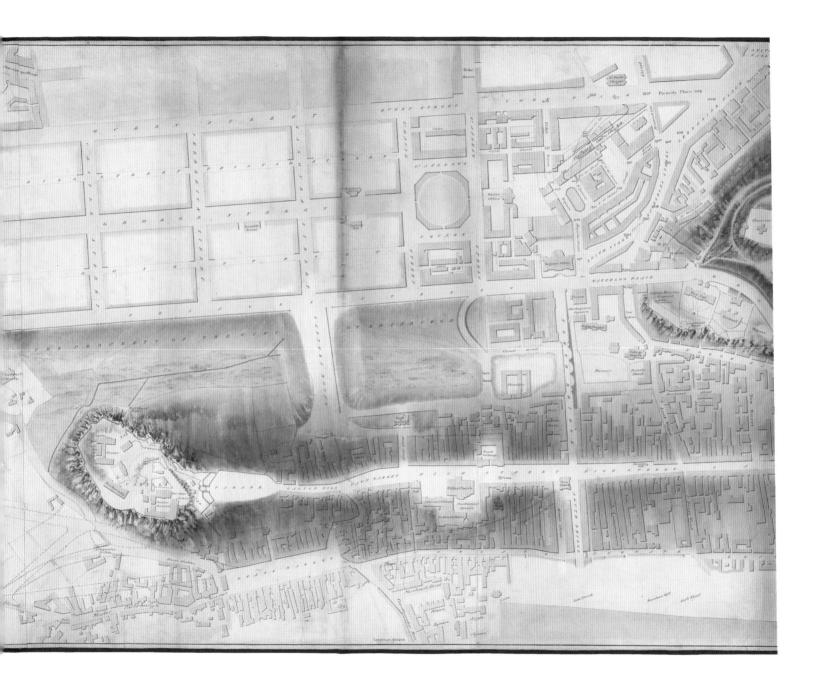

such a career in terms of socioeconomic background and connections. He is often acknowledged as a founding father of the civil engineering profession, but it is seldom appreciated how much the emergence of the said profession was a 'civilisation' of pursuits that had been, until the late Georgian period, largely the province of the military.

1822a

*Disentangling past, present and future
in Scotland's great county atlas*

At one level, this map is of interest as an overview of the Edinburgh environs in the early nineteenth century. But if examined in further detail, especially for its imaginary and non-existent features, it reveals a great deal about the practicalities of mapmaking at this time and the mixed, earlier provenance of the information used. It was compiled for John Thomson's *Atlas of Scotland*, which was one of the most ambitious county atlases of Scotland by an Edinburgh-based publisher, as well as the cause of one of the most spectacular map-publishing bankruptcies.

Thomson was born in 1777, the son of a merchant, and was a bookseller in Hunter Square by 1807, the same year he was admitted as a burgess of the city. He initially published various travel guides, moving on to atlases, including *A New General Atlas* (1817), the *Cabinet Atlas* (1819) and the *Edinburgh School Atlas* (1820). He advertised subscriptions for his *Atlas of Scotland* project in 1818, hoping the results would be available within a year, but quickly realised the immense difficulties of the project. Most available county maps of Scotland were out of date, often by decades, and even simply updating them would have been expensive and time-consuming. The costs of drafting and engraving entirely new maps for all the counties were even more daunting; just to break even, the project would have required a larger readership than Thomson had ever found in the past. Joining a growing trend among publishers, he moved to the fashionable New Town in 1824, just a year before he went bankrupt for the first time amid the general postwar economic downturn. Moreover, the *Atlas* project's cash flow failed to improve, and when Thomson finally completed it in 1832, his debts had grown to a massive £14,453, as against assets of only £9,969. Worst of all, the new, super-fine engraving techniques used in the *Atlas*, combined with a proliferation of new features and place-names on the ground, made the finished product look

John Thomson, *Northern Part of Edinburghshire* (1822)

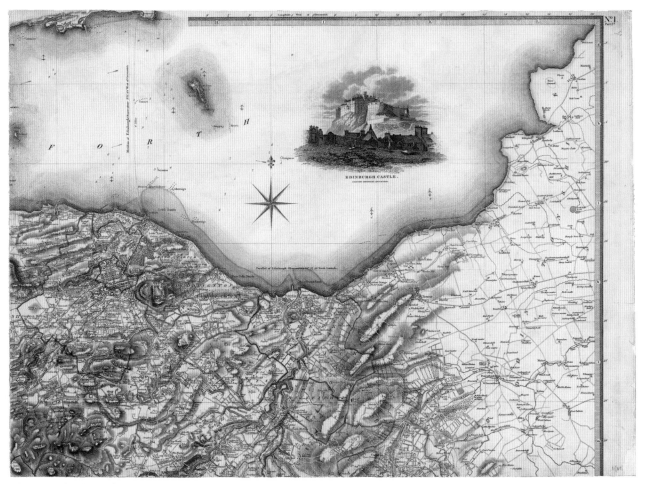

The top-right sheet of Thomson's map of the county of Edinburghshire, from which the details on pp. 120 and 123 are taken.

busy, murky and unlovable in comparison to almost all prior maps of the area, rendering the possibility of gentry wall-display a virtual non-starter. Perhaps the kindest thing that can be said about it, aesthetically speaking, is that it was precociously Victorian. Thomson's dogged attempts to purchase stock from creditors and unpaid engravers ultimately failed, and in January 1836 his entire stock of atlases and copper plates was sold by public roup at the Royal Exchange, where the rival firm of W. & A.K. Johnston acquired them.

Against this background, we can make more sense of this map. Most of the detail we are looking at is from a county map of Edinburghshire that had been surveyed by James Knox in 1812, engraved by Neele in London and published in 1816. It was considered by contemporaries to be a fine county map, in a technical sense: it was on a trigonometric base, at a good scale of 1.5 inches to the mile, included a variety of topographic details (especially in rural areas) and delimited parish boundaries. Knox had also been keen to flatter the gentry as possible purchasers, so gave the names of significant property owners; he also tried to ensure the map would remain up-to-date for as long as possible by anticipating certain planned developments.

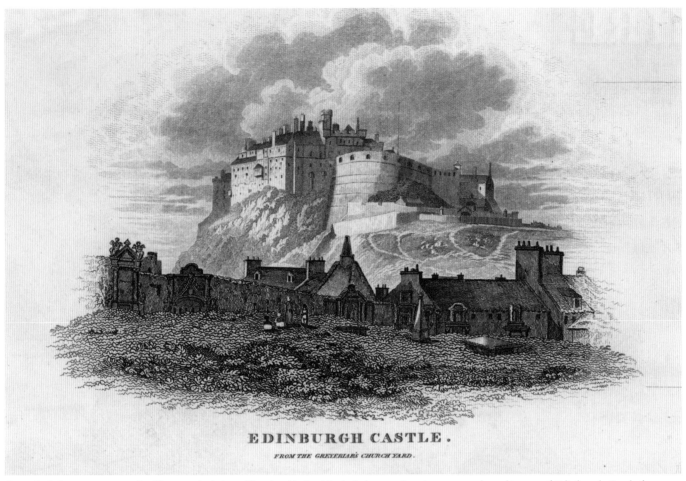

Several of the county maps for Thomson's *Atlas of Scotland* (1832) included attractive vignettes, such as this one of Edinburgh Castle from Greyfriars churchyard.

In revising Knox's map, Thomson relied on the assistance of 11 gentlemen who 'attested' to its authority. These included prominent civil engineers, road surveyors and landowners. Like Knox, this group also knew of a number of planned developments which they in turn added to the map. Perhaps the largest of these, covering some 250 acres, was the Calton New Town proposed for the east side of Leith Walk in 1819 by leading Scottish architect W.H. Playfair (1790–1857). Although building had started promptly, a number of difficulties stalled development until the 1860s, by which time it was executed to quite a different pattern – as might be expected, since Playfair was already dead and the Victorian era, with its very different architectural and planning priorities, was in full swing. On the other side of Leith Walk is another proposed but unbuilt estate at Pilrig. Further north, only two of the four wet docks at Leith were ever constructed, though, in fairness, both of these had been included on Knox's map too. The Deanhaugh estate to the west of Stockbridge and Coates estate in the West End also both show more buildings, in different patterns, than was really the case on the ground at this time.

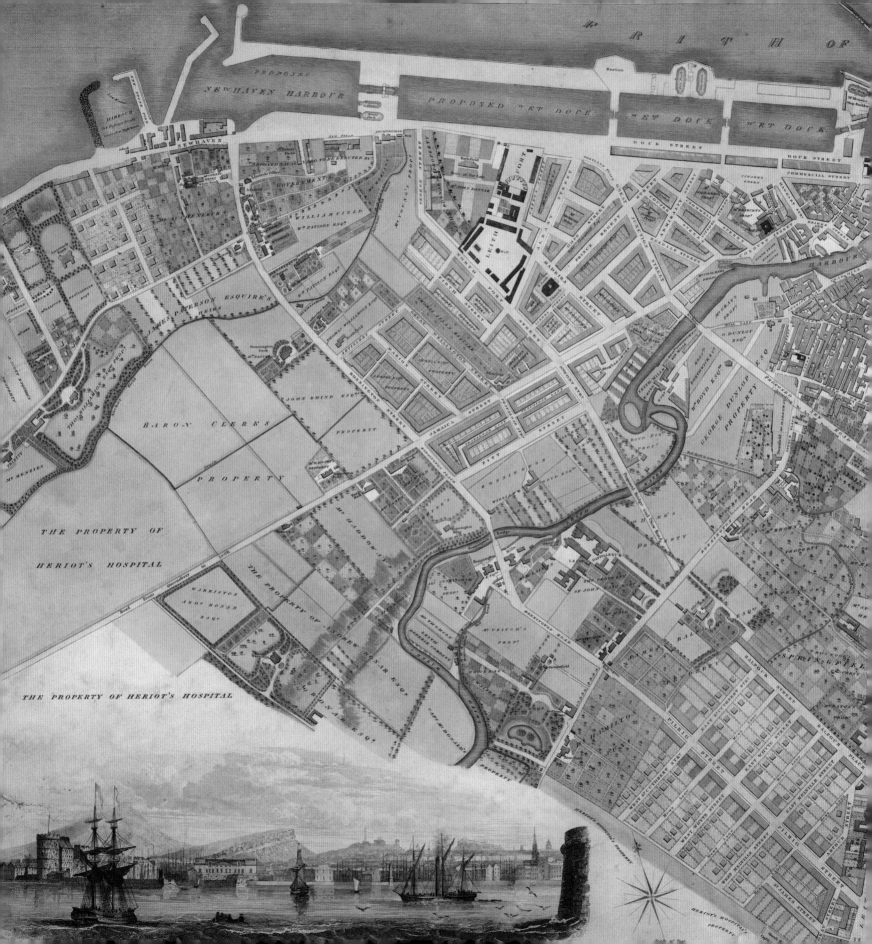

1822b

Leith at the time of King George IV's visit

This map was engraved and published in February 1822, perhaps as part of the build-up to King George IV's planned visit in August. On the day, the sedate harbour depicted here turned out to be rather chaotic, as the *Royal George* yacht, accompanied by steam packets and a large royal flotilla, was surrounded by welcoming steam vessels filled with cheering passengers. Owing to torrential rain, which also smothered a ceremonial fire on Calton Hill, the decision was made to delay the landing until the following day. Nevertheless, and in spite of the King's unpopularity and widespread radical opposition, Sir Walter Scott used the first visit of a reigning monarch to Scotland since 1650 as a vehicle for national unity, orchestrating a splendid pageant mixing medieval symbolism and resurrected tartanry. These 'one and twenty daft days' of ceremonial salutes, cannon fire, processions and parades began in Leith, proceeding up Leith Walk through a theatrical gateway where the king was presented with the keys to the city. As often happened in those days, most of the surviving structures commemorating his visit were built considerably later: the first stone of George IV Bridge was not laid until 1827, whilst his statue at the junction of George Street and Hanover Street dates from 1831.

In the view, from our vantage point by the town's martello tower [1815] in the right foreground, we can see the Signal Tower to the left on the corner of the Shore and Tower Street. Converted from its former use as a windmill as recently as 1805 and fitted out with new battlements, this used flags to communicate the depth of water in the harbour to incoming ships. The large rectangular building at centre left is the new Custom House, built in 1812, whilst on the far right can be seen the elegant spire of North Leith Church, constructed 1814. The distinctive outlines of Salisbury Crags, Calton Hill

Charles Thomson, *Plan of the Town of Leith and its Environs with its Intended Improvements* (1822)

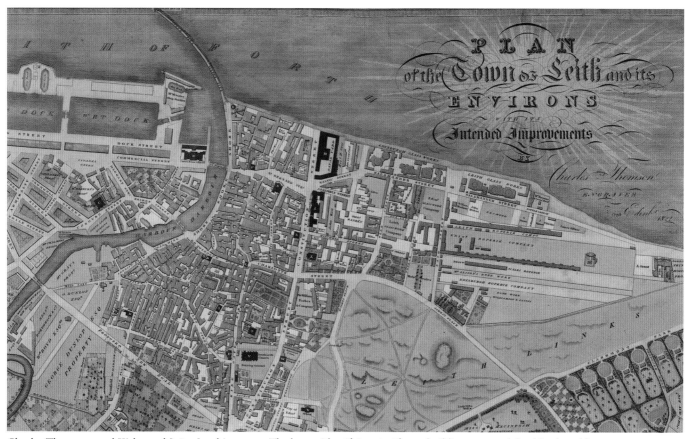

Charles Thomson used Kirkwood [1817] as his source. The letters identifying significant buildings are explained in the table on p. 127.

and Edinburgh Castle form an attractive distant horizon.

As could be surmised from these construction dates, Leith was entering a period of major expansion, following the completion of John Rennie's successful 1799 scheme for a new pier on the eastern side to improve the scour of the Water of Leith, and two large wet-docks to the west of the harbour. The first of these was complete by 1806, and the second by 1817, along with three dry-docks and new drawbridges. The massive costs of £285,000 for the whole scheme by this time led to its temporary suspension, and the proposed additional wet-docks shown as extending west to Newhaven were never in fact constructed. Instead, reclamation from the 1830s onwards proceeded north of these docks, so that by the 1930s, they would lie more than 1,300 yards inland of the harbour breakwaters; and the 1809 martello tower, seen in this view with waves lapping up its sides, would be landlocked. Trade was significant and growing: in 1826 the customs duties collected in Leith totalled about £480,000 per annum, with 3,628 vessels arriving. Many of these were coastal, with 21 smacks sailing three times a week, and four steam vessels twice a week to London; but other boats traded with Russia, Denmark, Sweden, Norway, Hamburg, Holland, France and Spain, as well as further afield, including the West Indies and America.

Relations between Leith and Edinburgh were frosty at the best of times, a situation not helped by the City of Edinburgh's historic ownership of the port of Leith and its revenues. Nevertheless, following the bankruptcy of the city in 1833, an Act of Parliament in 1838 formally transferred the Leith dock customs to a new Dock Commission, made up of representatives from Leith, Edinburgh and the Treasury. Additional political independence came through the Town Council Act of 1827, which allowed Leith to elect its own council and magistrates, with even greater rights added via acts of 1833 and 1838. The formal re-appropriation of Leith by Edinburgh in 1920 would meet with bitter local opposition.

Thomson's map also allows us to see several of the industries related to the port: shipyards, rope-works ('roperies'), mills and distilleries, and new banks facilitating trade, as well as the significant glass and bottle works to the east of the harbour, dating back to the early eighteenth century. The map also shows partially executed schemes for significant new thoroughfares and public buildings, in clear contrast to the dense tangle of streets forming medieval Leith. Constitution Street had been gradually laid out from the 1790s, but appears here close to its present-day form. Great Junction Street had been planned around 1800 as a grand new road, largely following the ramparts and ditches of the mid-sixteenth-century fortifications [1560], to connect the Edinburgh–Leith road at the foot of Leith Walk to the new proposed docks. Its northern section was completed first, including the Junction Bridge of 1818, but further south it was not actually completed until the 1830s.

Significant new public buildings included the new Assembly Rooms on Constitution Street (1809–10), the new markets in 1818 on the site of the Old Custom House and Excise Office, and the new Trinity House in Kirkgate from 1816, on the foundations of the previous building of 1555. The former Trinity House had been built by the Incorporation of Masters and Mariners, a fourteenth-century institution that dispensed aid to poor or infirm sailors and, by the 1630s, collected money for the Forth's primitive coal-fired navigational aids. From 1797, it gained new responsibilities for the licensing of pilots. However, its similarity of name to the lighthouse authority serving England, Wales and Gibraltar [1693] is a coincidence, rooted in both bodies' origin in the pre-Reformation heyday of guilds. From 1786, lighthouses in Scotland and the Isle of Man would become (and remain) the responsibility of the Northern Lighthouse Board, headquartered at 84 George Street in the New Town.

Charles Thomson (fl. 1818–1833) was a highly skilled engraver, with premises on the High Street in Edinburgh. He also engraved several other maps of Edinburgh in the 1820s, including the one by Brown & Wood included in this volume [1823]. He was heavily indebted to Robert Kirkwood's fine and detailed survey of 1817 for his source information: hence the inclusion of property owners, and the similar style used to indicate proposed streets and buildings, particularly to the north and west. Thomson even included the same key to buildings:

A. Leith Port	O. Council Chamber
B. North Leith Church	P. Tolbooth
C. Third Burgher Meeting House	Q. Charity School
D. Custom House and Excise	R. Church of John Knox
E. Signal Tower	S. Second Burgher Meeting House
F. British Linen Company Bank	
G. Leith Bank	T. First Burgher Meeting House
H. Naval Yard	
I. Assembly Rooms and Post Office	U. Trinity House
K. Chapel of Ease	V. South Leith Church
L. Water Reservoir	W. St James's Chapel
M. New Flesh Market	X. King James VI Hospital
N. Fish and Poultry Markets	Y. High School

In spite of eighteenth-century copyright legislation, the costs and delays associated with going to court encouraged many engravers to take the risk of plagiarism until the later nineteenth century. Thomson's map does include a few new updated details, particularly of buildings and landowners, and excludes Kirkwood's proposed canals, possibly because the end-point of the Union Canal had recently been confirmed as Port Hopetoun. Thomson issued a revised version of this map in 1827, with a new, respectful scene depicting George IV's actual arrival.

1823

Announcing the Greek Revival

'This magnificent City, the Metropolis of Scotland . . . has been justly called the Modern Athens . . . and "City of Palaces"', John Wood wrote of Edinburgh in his *Town Atlas of Scotland* (1828), in which this map was included. One of the most important private surveyors in this period, Wood published more than 140 plans of British towns between 1818 and 1845, of which 57 were Scottish. Most were based on original surveys by Wood himself, though this one was a revision of the 1820 plan published by Thomas Brown at 1 North Bridge, and engraved again by Charles Thomson, who had also engraved a map of Leith in the previous year [1822b].

Looking at this plan from the top down, we see proposed developments including streets and buildings on the Dean Estate; a rectangular layout for Saxe-Coburg Place; the Playfair New Town on either side of Leith Walk and north of Calton Hill; the National Monument (identified here as 'National Ch[urch]'); the projected streets in the Orchardfield Estate east of Lothian Road; and Castle Terrace. What would become Johnston Terrace is sketched in, although not in the correct alignment with King's Bridge, whilst Victoria Street and George IV Bridge are shown at least a decade before their construction. North of Old College, a proposed street anticipates the Chambers Street development by half a century.

Particularly owing to the financial crash of 1825–26 and arrested development thereafter, building took quite a different pattern from that depicted, often decades later. The National Monument, to Scotland's war dead of the Napoleonic Wars, was designed by W.H. Playfair and C.R. Cockerell as an exact replica of the Parthenon in Athens; however, construction was

---

Thomas Brown and John Wood, *Plan of the City of Edinburgh, Including all the Latest and Intended Improvements . . .* (1823)

abandoned after just three years (in 1829) because of lack of funds. An equally important exemplar of the 'Greek Revival' style was the new building for the Royal High School, also noted by Wood as under construction. Various topographical and intellectual similarities between eighteenth-century Edinburgh/Leith and classical Athens/Piraeus had been widely noted, but the main thing that quite literally cemented Edinburgh's status as the 'Athens of the North' was its wholesale adoption of retro-Athenian temple architecture – particularly by Playfair, who also designed the Royal Scottish Academy, the Scottish National Gallery, and the City Observatory (shown here as 'Scientific Observatory', its name from 1818 to 1822).

Whilst keen to flatter the inhabitants for their benevolence

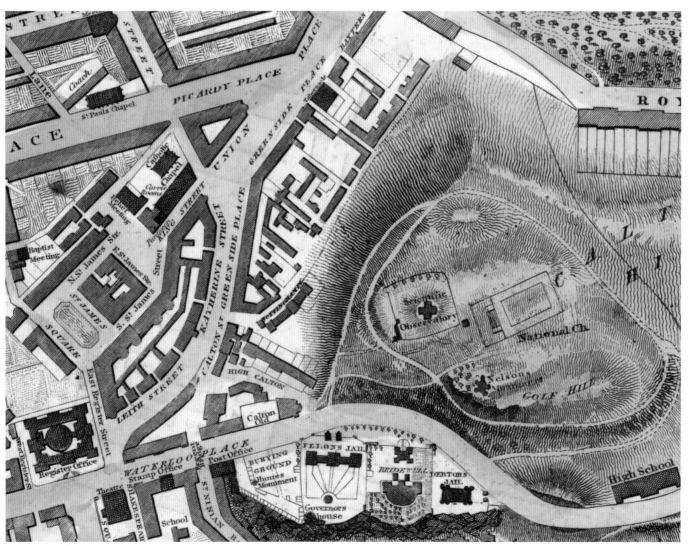

This detail of St James Square and the west of Calton Hill shows Waterloo Bridge and Regent Terrace, the City Observatory and the National Monument (originally designed as a church), all of which had been constructed within the previous decade.

– 'in no city are charitable institutions more numerous . . . for the alleviation, or cure of every form of human misery' – Wood also noted the dreadful, overcrowded slums that were the source of so much of the demand for charity. 'The lands, or houses in the wynds or lanes on the declivities on each side of [the High] Street, are very high; these lands have a common stair, and it is not uncommon to find from 18 to 24 families in the same building; thus rendering these crowded abodes, not only unhealthy and uncomfortable, but dangerous from fire.'

Wood is thought to have trained in Yorkshire, but married in St Cuthbert's Church, Edinburgh, in 1811, and made Scotland his permanent base after acquiring Canaan Grove, a grand house near Morningside, in 1813. He died 'from cramp of the stomach' in Portobello in 1847.

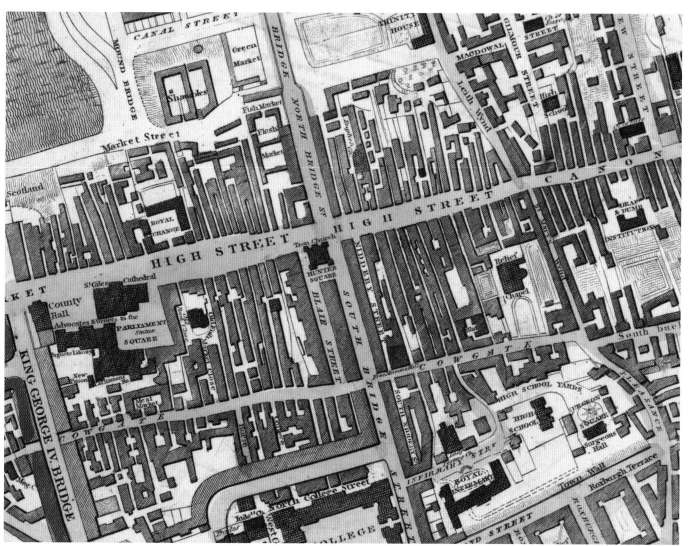

This detail of central Edinburgh anticipates several new developments. George IV Bridge was constructed as shown by 1834, but not the proposed street North of Old College.

1826a

Parkland, estate mapping and suburbia

In the century between 1750 and 1850, private land surveyors supplanted military engineers as the dominant group in mapmaking in Scotland. Usually employed by landowners to survey their estates and gardens, many became expert planners of improvements in agriculture such as drainage, enclosure, and new crops and rotations, as well as designers of new gardens and landscapes. Before the Ordnance Survey's mapping of Scotland from the 1840s to the 1880s, large-scale maps from an original private survey, such as the one shown here, were essential for estate management. This was due especially to their accurate measurements of acreages, allowing rents to be calculated. Estate maps were usually working documents, and this one is typical in having various annotations: of revised field boundaries, carriage drives and acreages updated from the medieval system of ploughgates, oxgangs, Scots acres, rods and falls, with the last-named three designated 'A', 'R' and 'F' in the key at lower right.

Very much part of this tradition, this hand-drawn map by David Crawford (d. 1829) is a useful reminder of the importance of the productive agricultural hinterland which Edinburgh was increasingly built around, or over, in the later nineteenth and twentieth centuries. The Keith family hailed from Kincardineshire, but acquired the estate of Ravelston in 1726. Their attempts to gentrify the workaday place-names of the area yielded some amusing juxtapositions: Barnyard Park, Nineteen Rigs Park, and West Bog Park perhaps foremost among them. 'Ravelstone Tower', the ruin of a Z-plan tower-house built in 1622 for previous owner Sir George Foulis, can be seen nestling in trees to the far left.

Alexander Keith (1737–1819) – an uncle of the Alexander Keith who owned the estate in 1826 when this map was made – undertook archaeological excavations at Ravelston and read papers on his finds to the Society of Antiquaries. He oversaw the construction of Ravelston House, a late-Adam-style

---

David Crawford, *Plan of the Estates of Ravelston and Corstorphinhill* (1826)

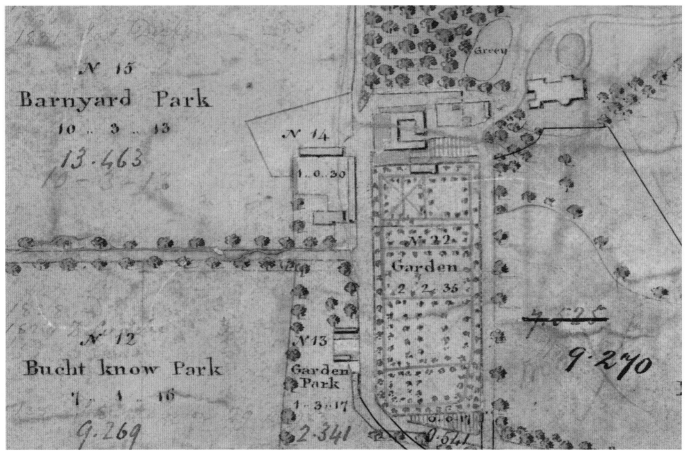

The formal gardens and tree-lined avenues leading to Ravelston House, today part of The Mary Erskine School.

Palladian villa, in the 1790s. He also supported John Thomson's great atlases [1832], and Thomson's *A New General Atlas* (1817) contained a dedication to Keith featuring his coat of arms. The younger Alexander, who acted as knight-marshal during the legendary visit of George IV to Edinburgh [1822b], subscribed to Thomson's *Atlas of Scotland* (1832).

Ravelston lies only two miles west of Edinburgh Castle, but in spite of this and the proximity of the Queensferry Road and Blackhall (p. 135, upper right), the fields and woods shown on this map can largely still be identified today. In the post-Second World War period the expanding suburban villas of Blackhall, Murrayfield, Corstorphine and Clermiston encroached on all sides, but came no further. This largely reflects the importance of the area for golfers: the estate lands were originally leased in the late nineteenth century (and later sold) to form Ravelston Golf Club to the north, and Murrayfield Golf Club to the south, while the Edinburgh Zoological Park acquired an area to the southwest in 1912. The house and remainder of the estate were developed from 1964 to form part of Mary Erskine School. This school was established as the Merchant Maiden Hospital in 1694 by the same Mary Erskine who founded the Trades Maiden Hospital ten years later [1742], despite the fact that the Trades (incorporated 1562) and the Merchant Company (1681) were major rivals for political power within the city. The driveway shown at the right of this map, curving at the end below the gardens,

1826

is now in public use under the name Ravelston Dykes Road.

Crawford, who was living at 1 West Nicolson Street in 1820, was born into a distinguished family of land surveyors based in Edinburgh. His father William (fl. 1774–1828) had trained with John Home, mapping Assynt in the 1770s, and went on to work with John Ainslie. William had another land-surveyor son, also William, who was active until 1845 and well known for his county mapping of 1839–45 in partnership with William Brooke.

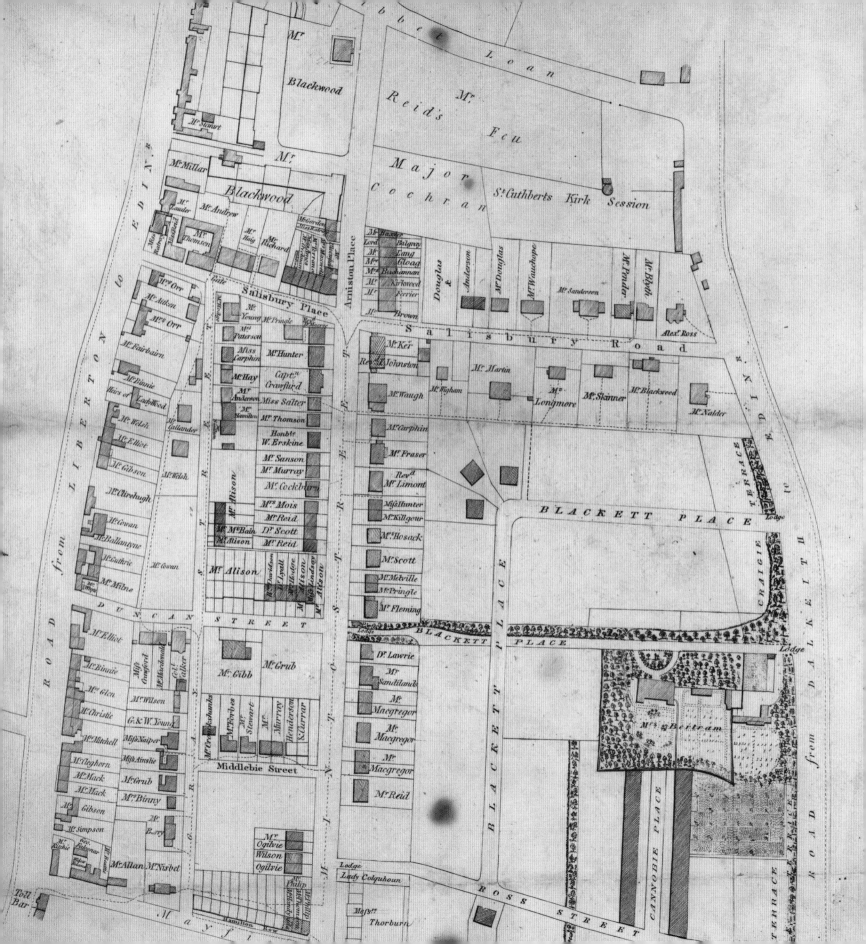

1826b

An early 'gated community': Newington

This plain but detailed plan shows the rapid development of a part of Newington planned very much as an upper-middle-class residential estate, and helpfully names all the property owners.

The sensible north–south alignments and widths of Causewayside, part of Newington Road and Dalkeith Road were in fact originally suggested in a feuing plan for part of the Easter Muir in 1586. However, owing to its gentle south-facing slopes, good soil and distance from the Old Town, the area remained largely arable until the early nineteenth century. Nevertheless, traffic increased, particularly in the later eighteenth century – witness the 'Toll Bar' and toll-taker's cottage with its distinctive protruding observation window, both at lower left. What would become Minto Street was authorised by an Act of Parliament in 1794, and was labelled by Ainslie [1804] as the 'New Intended Road to London'.

In 1803 the whole area was acquired by Benjamin Bell, a celebrated surgeon and the great-grandfather of Dr Joseph Bell, prototype of Conan Doyle's character Sherlock Holmes. Several of the streets were named in connection with the Bell family and their friends. Blackett House in Middlebie parish in Dumfriesshire was the family home, and much of the development of the Blacket Estate (as it is now spelled) was just beginning at the time this map was drawn. Duncan Street was named after Admiral Adam Duncan, celebrated victor over the Dutch fleet at the Battle of Camperdown (fought off the coast of North Holland in 1797), who was one of Bell's friends and patients, whilst Arniston Place (upper left) was named after the Dundas of Arniston family, also friends of Bell's and neighbours to the Duncans in George Square. In 1805 work began on Newington House, the large, detached house towards the southeast corner of the district, set in attractive landscaped grounds. Although Bell died the following year, probably before the completion of the house, the ongoing

John Leslie, *Plan of the Lands of Newington and Belleville* (1826)

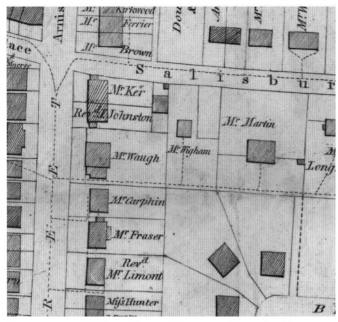

The junction of Salisbury Place and Causewayside retained its residential character initially, but new manufacturing gradually encroached on it. From 1870, the Middlemass biscuit factory occupied this site, and since 1988, it has been home to the National Library of Scotland's Causewayside Building.

development of the area continued under his son George, also a surgeon.

Comparison with Kirkwood [1817] confirms a spate of building in the intervening nine years, particularly along Minto Street, Upper Gray Street, the north side of Middleby Street and Duncan Street. George Bell commissioned James Gillespie Graham (1776–1855), who had been responsible for planning Moray Place in the New Town, to draw up the plans for Blacket, whose feu charter forbade 'any manufactory of soot or blood, breweries, distilleries, tan works, kilns or any manufactory which could be regarded as a nuisance'. The area would become, in effect, an early 'gated community' of mostly large detached and semi-detached villas, and Gillespie Graham's neo-Tudor gate piers and porters' lodges, controlling access to Newington House, can be seen on this map at both ends of Blackett Place.

All the houses shown here were quite substantial, with at least one live-in servant, and were typically occupied by professionals, businessmen and those with independent means. However, west of Minto Street, much of the development halted or slowed soon after this time, and, with looser feuing conditions, gap sites were often occupied by small businesses, which remain a major focus of the area at the present time. By the 1870s, there was a printing and publishing works, a laundry, a biscuit manufactory and a 'horse bazaar' between Causewayside and Upper Gray Street; to what extent any of these were considered a 'nuisance' was very much in the eye, and nose, of the beholder.

The map also names a number of more famous inhabitants, including William Blackwood, publisher and founder of *Blackwood's Edinburgh Magazine*, who moved to the eastern end of Salisbury Road in 1806. The property was owned by Blackwood's descendants until 1857, and survives today as the Salisbury Centre. Further west, at the corner of Salisbury Place and Minto Street, was the home of the 'Revd Dr Macree', author of the *Life of John Knox* (first printed in 1812), one of Blackwood's most successful early publications. In between these two houses we find Mr Longmore, whose grandson, John Alexander Longmore of Deanhaugh, would found the Longmore Hospital for Incurables in 1880 on the north side of Salisbury Place, replacing several former houses there. The hospital closed in 1991, but the building survives today as the headquarters of Historic Environment Scotland, an executive agency of the Scottish Government. The most famous cartographic residents of this area would arrive in 1911, to occupy the newly constructed Edinburgh Geographical Institute building on the south side of Duncan Street, on the land shown here as owned by Mr Gibb.

Earlier feuing plans of 1795 and 1806 confirm a consistent vision for the area, but some flexibility in its implementation, with development proceeding along major routeways from west to east. The map was surveyed by John Leslie, a West Port-based estate surveyor also known today for his mapping of Edinburgh roads, and for the reorganisation of Cramond village in the 1820s.

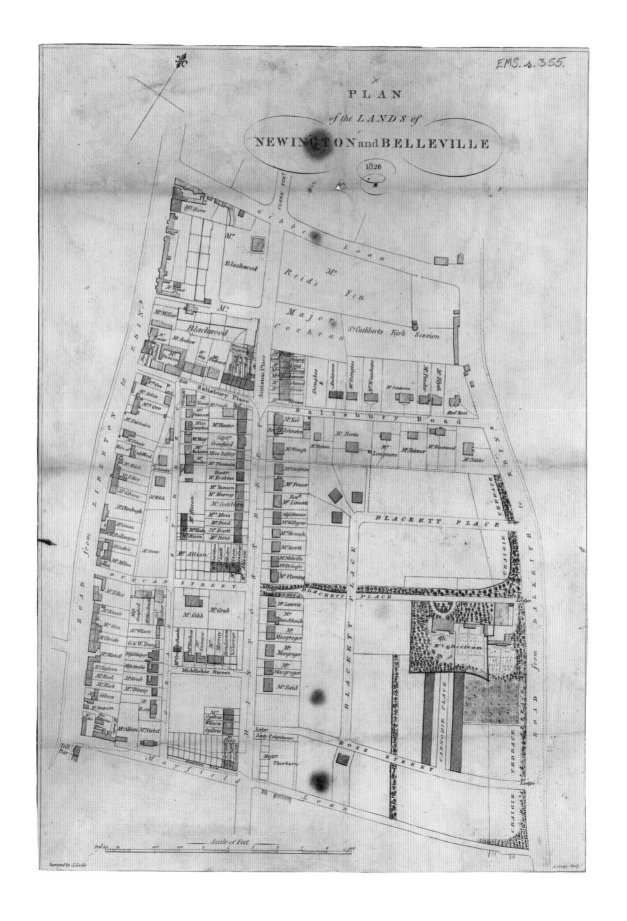

1832

Extending the franchise on paper

This map – one of a set of 75 Scottish town plans – shows in red the new Parliamentary boundaries that came into effect through the Great Reform Act of 1832. This marked a significant shift in power relations, away from aristocratic interests and county elites to the growing urban middle classes and tradesmen. The Scottish electorate expanded 16-fold, from 4,239 in the 1820s to more than 65,000: a much more dramatic leap than the 62 per cent increase in the UK electorate as a whole. The overall number of Scottish Members of Parliament rose by eight, to 53, and this entire increase was in burgh rather than county representation. Edinburgh was allowed a second MP, whilst some older royal burghs such as Aberdeen, Dundee and Perth were given a dedicated MP for the first time. (Previously, 14 groups of four to five royal burghs had each shared a single MP, and Edinburgh alone was not in any such group.) The wave of revolutions that would rock more than 50 countries in Continental Europe and Latin America in 1848 had broadly similar electoral reforms as their main aim; most would fail.

None of this is to suggest, however, that late Georgian Britain had achieved a state of democracy. The extension of the vote to owners of property worth £10 included only 10,607 houses in Edinburgh, out of an 1831 population of 136,294. Wide swathes of the city still remained without any voters – only one in eight adult males was eligible to vote even after 1832 – and, in the absence of a secret ballot, traditional interest groups still held sway. The later Reform Acts of 1867 and 1884 would go much further in extending the franchise, yet many unskilled labourers and all women were not able to vote until 1918. The sharing of MPs by groups of smaller

---

William Murray and J.W. Pringle / House of Commons, *Plan of Edinburgh and Leith*,
from *Reports upon the Boundaries of the several Cities, Burghs, and Towns in Scotland,
in respect of Members to serve in Parliament* (1832)

burghs also persisted, with Haddington, North Berwick, Dunbar, Lauder and Jedburgh having a joint MP until 1885, and Leith, Musselburgh and Portobello sharing one until the end of the First World War.

The passage of the Representation of the People (Scotland) Act 1832 was nevertheless welcomed by those seeking a fairer franchise, and resulted in a celebratory procession of more than 15,000 people through the streets of Edinburgh. Perhaps five or six times this number also took time off to witness this and to join in the rejoicing. On the whole, the public demonstrations around Reform were peaceful and patriotic, but not always: one procession into the Tory heartlands of the New Town ended with violence and broken windows.

The burghs were visited by two commissioners, Mr William Murray and Captain J.W. Pringle of the Royal Engineers, who drew up plans to illustrate the law's verbal descriptions of the new constituency boundaries. This was very much a rush job: according to their report, 'all the burghs were visited twice, some of them oftener', meaning that more than 150 visits were made in less than 120 days. Unsurprisingly, given these circumstances and their primary purpose, the plans provide only general detail for the central urban area, with buildings clumped together as shaded blocks – though public buildings are shown in a fair degree of detail. In this context, it is worth comparing this map's vast square block of Register House as completed, against the smaller, incomplete structure depicted by Kirkwood [1819a]. Major streets, tolls, canals, bridges and quarries are also shown, along with major industrial units including the Gas Works at Tanfield Hall, Canonmills – a curious structure, variously described as Italianate, Prussian or 'Moorish' in appearance, which was fully operational from 1825. By 1820, gas for street lighting had already been adopted in London, Paris, St Petersburg and Baltimore in Maryland.

The first General Assembly of the Free Church of Scotland would meet in one of the buildings of the Tanfield Gas Works complex on 18 May 1843.

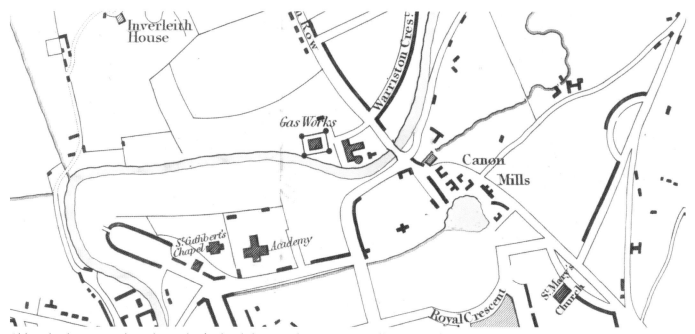

Although schematic, and not the result of a detailed survey, this map is a rare illustration of actual rather than projected buildings and streets: usefully showing the new Tanfield oil and gas works, and new houses along Brandon Street and Warriston Crescent.

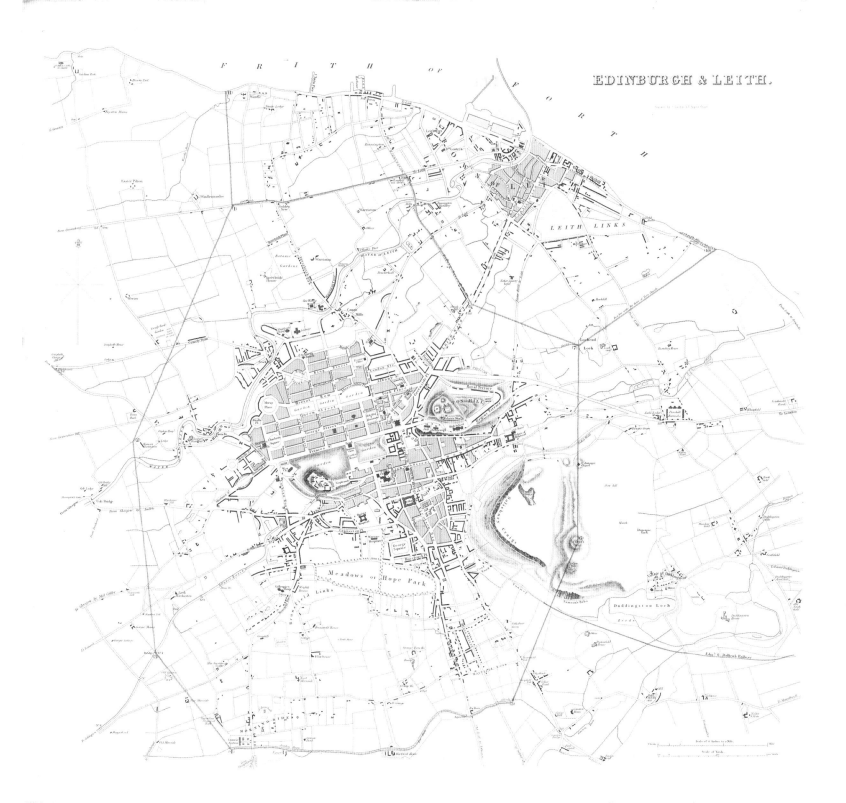

1834a

Maps for 'the masses'

The Society for the Diffusion of Useful Knowledge (SDUK) was conceived in 1825 by Henry Brougham (1778–1868), a powerful Liberal politician who had grown up in shabby-genteel circumstances in the Edinburgh of the Scottish Enlightenment, and who was equally at ease writing legal briefs, popular journalism and scientific treatises on optics. Over the next 20 years the SDUK were responsible for an extensive non-profit publication programme, focusing on the ideal of making available cheap but authoritative printed material for the 'man in the street' (see also [1784]). In spite of the steady growth in literacy, in travel, and in knowledge of geographical subjects, maps were still relatively expensive, and SDUK proposed a low-budget map publication programme as an adjunct to their Library of Useful Knowledge. This would lead directly to the publication of more than 3 million maps, a number that probably rivalled the entire map output of the island of Britain since the invention of printing in the fifteenth century.

Henry Bellenden Ker (d. 1871), a Lincoln's Inn colleague of Brougham's, pushed the map project forward and enlisted the help of Captain Francis Beaufort (1774–1857) of County Meath, Ireland – an accomplished surveyor, and soon to become Hydrographer of the Admiralty. The son of a respected topographer/clergyman of French Protestant descent, Beaufort was very exacting, which often led to delays in the publication programme; he enlisted the help of J. & C. Walker, also highly regarded for their high standards, as engravers.

Nevertheless, by the early 1830s the SDUK map series exceeded 200 titles. Of these, 51 were city plans (including this one), which tend to be the most highly valued today. They were largely the work of W.B. Clarke, an architect whose commitment and enthusiasm matched Beaufort's. Whilst the

---

SDUK, *Edinburgh: Reduced under the Superintendence of the Society for the Diffusion of Useful Knowledge with the Permission of Messrs. Laing and Forbes from their Large Plan* (1834)

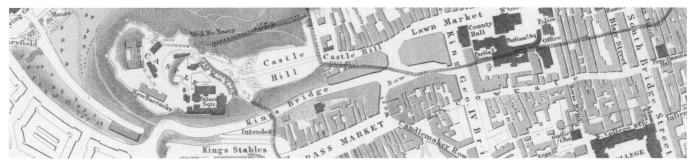

As well as the position of the town walls, shown in red, this SDUK map illustrates an unrealised proposal for the Western Approach to the Lawnmarket running to meet the West Bow midway through its ascent.

SDUK managed the authorship, design and engraving of the maps, all other costs and materials were the responsibility of the firm of Baldwin & Cradock, who published the maps serially in instalments of two maps each. The present map was paired with an attractive three-part map of Scotland at 12 inches to the mile. These publications were subsidised by the SDUK to keep the price down to one shilling per instalment.

As well as being finely engraved, with an attractive view of Edinburgh Castle from the south and outline pictures of significant public buildings along the lower margin, the SDUK map was coloured with a light blue wash for water and green for gardens and open spaces. This detail from the South Side presents a number of interesting proposed developments as well as a relatively up-to-date street layout. George IV Bridge had been authorised in 1827, and was officially completed in 1834, the year this map was published; and the proposed Forrest Road, continuing to Middle Meadow walk, is also shown. A curious feature not found on many other maps is the Western Approach to the Lawnmarket from King's Stables Road, authorised by an Act of Parliament of 1827, running to meet the West Bow midway through its ascent. In fact, the original road that would become Johnston Terrace, running up to the Lawnmarket, was not opened until 1836 [1836]. Victoria Street, leading directly from the Grassmarket up to George IV Bridge, had been planned in the year this map was published and was complete by 1837. The positions of the historic town walls – the King's Wall of c.1420–50, the Flodden Wall of 1514–60 and Telfer's Wall of 1628–36 – are prominently shown with a red line.

The SDUK map also shows the confused and stalled building developments on the Orchardfield and Drumdryan Estates by the 1830s. Orchardfield, comprising a rectangular wedge of land running eastward between Lothian Road and Castle Terrace, suffered from legal complications, uncertainties over the Town Council's proposed transport improvements, and the general slump in the property market. Further south, the Drumdryan Estate of James Home Rigg had likewise suffered in the 1820s from uncertainties over the street layout and restrictions in feuing contracts, which in the depressed property market resulted in disorganised and piecemeal development. Brougham Street, Rig Street and the attractive Morton Crescent, bordering the Meadows as shown here, were never actively constructed along these lines. The estate was only really developed after its sale in the 1860s to the Edinburgh builder James Steel.

This map saw a number of later iterations, partly reflecting the successive publishers of the SDUK maps. Baldwin & Cradock became insolvent, and so SDUK took on publishing itself for several years from 1838, before handing it on to Chapman & Hall in 1842. The SDUK map plates were eventually acquired by various publishers, including Edward Stanford in 1856, and reissued in 1863–65. In 1853 the SDUK Edinburgh map was reissued by J. Cox in London, with the red town walls replaced by the new routes of the railways – the Caledonian, North British and Edinburgh & Dundee – which had been constructed a few years earlier.

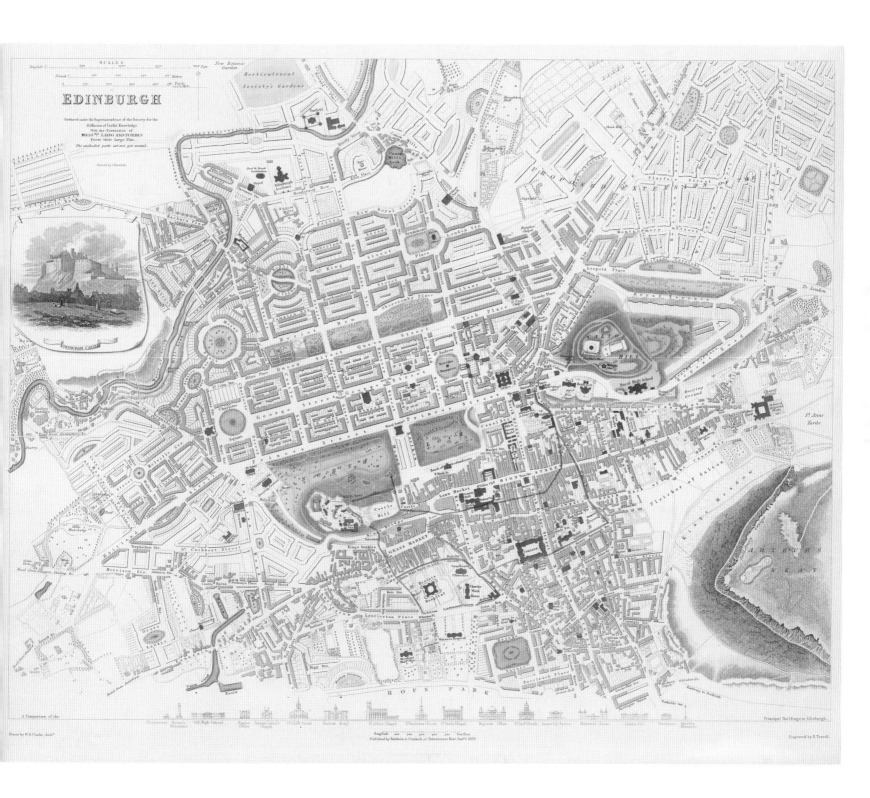

1834b

A new harbour for the Age of Steam

By 1833, steam power was having significant effects on Britain's maritime trade. While steamships would not fully supplant sail for another century, they were of ever-increasing importance – and size. The East and West Docks just to the north of Commercial Street in Leith were narrow and awkward to enter, charged high dues, and were unsusceptible to major improvement – particularly in the straitened financial circumstances that followed the great crash of 1825 (see [1836]). Nevertheless, there was a pressing and widely recognised need for a large, deep, modern harbour for Edinburgh capable of accommodating the largest of the new ships; unsurprisingly, a particularly prominent promoter of the idea was the General Steam Navigation Company, which had been founded in London in 1824. In the event it was the 5th Duke of Buccleuch who decided to have such an anchorage built, not as an improvement to an existing one, but on previously undeveloped coastal land adjoining his dramatic ogee-roofed stately home, Caroline Park (1685–96). The house, known as Royston until 1740, is shown here a short distance to the southwest of the proposed harbour works. The duke's search for a cheaper means of exporting the growing quantities of coal mined on his estates in Midlothian was an important motivation for the scheme.

Robert Stevenson, the leading marine engineer of his time (see [1819b]), was engaged by the duke to design Granton Harbour in 1834. A rival plan for a harbour at Trinity having been rejected by the House of Commons in 1835 (the light 'chain' pier shown was unsuitable for freight), Stevenson's plan was presented to Parliament in 1836 and received the assent of King William IV in 1837. In fact, construction had been going on for more than a year, since Buccleuch 'was satisfied that in view of the terms under which the original

Robert Stevenson, *Chart of the Firth of Forth from Queensferry to Inchkeith, Showing the Relative Position of the Proposed Harbour at Granton* (1834)

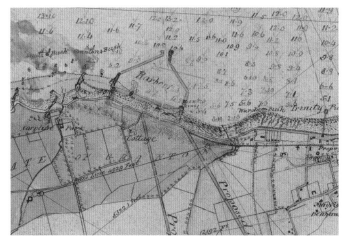

Both of these direct routes to Granton Harbour by rail and road were implemented, but on different lines to those shown here. The shaded blue area marks land owned by the Duke of Buccleuch.

grant of the lands had been made from the Crown, such sanction was unnecessary'. While the layout of the breakwaters shown on this map would change only slightly, the pier was revised fundamentally, to run perpendicular to the shoreline rather than parallel to it, as seen here. Eventually, Stevenson's two 3,000-foot breakwaters would flank a 1,700-foot north-to-south pier, constructed almost entirely of sandstone quarried on the Granton estate. This change to the plan would create two harbours: the Western of 67 acres (completed 1851), and the Eastern of 52 acres (1863). In all, construction would last for 27 years, 13 of them after Stevenson's death from old age – though the pier was officially declared open on Queen Victoria's coronation day in June 1838.

Stevenson's 'Proposed New Road' would have cut through Canonmills Distillery and the Botanic Gardens (located in Inverleith since 1823) and was never built. However, north of Ferry Road (running from Leith to the west), Granton Road was constructed by 1848 along a similar course to that shown here. Instead of laying new railways in from Fountainbridge in the west or from the Leith docks in the east, the Edinburgh, Leith & Newhaven Railway Company, incorporated in 1836, put in a new direct line from Scotland Street station to Trinity Crescent, to connect with the Trinity Chain Pier. Following storm damage, a new extension was completed in 1846, curving around to the west to Granton; and the following year, the Scotland Street tunnel [1847] extended the line to Canal Street. This would remain the main rail route to Granton until the 1868 completion of the new North British Railway line from Waverley via Abbeyhill and Bonnington.

Organised tourism – whether to Scotland from England, or to the Highlands from the Lowlands – was a relatively new phenomenon at this date, and was a major driver of the 'great increase in the number of passenger steamers plying to the Forth'. The queen herself arrived via Granton Harbour for her fateful first visit to Scotland, in 1842; but 'Victoria Jetty' as a name for the Middle Pier did not stick.

A correspondingly large harbour was built at Burntisland in Fife after 1841, as a joint venture between Buccleuch and Sir John Gladstone of Fasque, father of William Ewart Gladstone (four times prime minister between 1868 and 1894). However, the red lines connecting Trinity and Granton to Burntisland on this map clearly indicate that the importance of ferry services between these points predated harbour improvements on either side of the Forth.

The world's first public 'roll on/roll off' train-carrying ferry service, running between Granton and Burntisland, opened in 1850 and lasted for 40 years, i.e. until the opening of the Forth Bridge. Its first vessel, the 399-ton *Leviathan* (1849), essentially consisted of two sterns joined together, each with its own steam engine, funnel and paddle-wheel. In spite of or because of this unorthodox arrangement, the rail-ferry technology was highly successful. Its details were devised by the firm of Grainger & Miller, but the idea for it came from a very young Thomas Bouch (1822–1880), who is now remembered – if at all – for designing the Tay Bridge that collapsed with the loss of 75 lives on 28 December 1879.

It is one of many testaments to Robert Stevenson's skill as an engineer that, shortly before the Second World War, a centenary history of Granton Harbour noted it had so far 'required no maintenance or repairs whatsoever'!

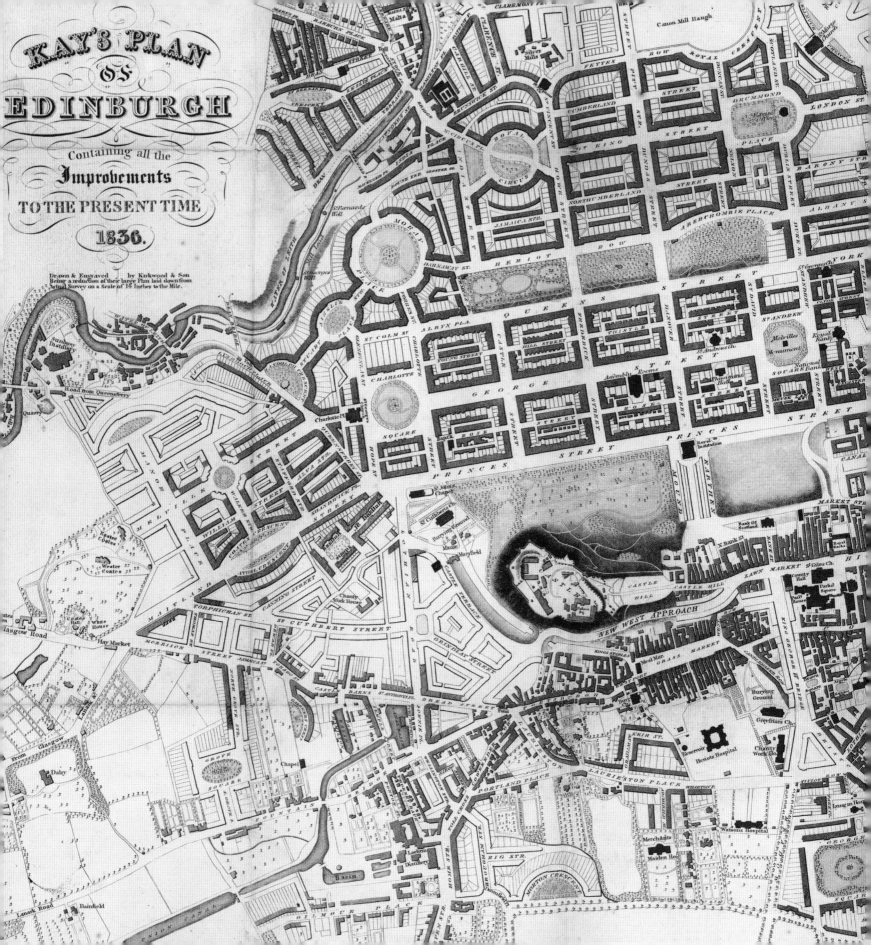

1836

Development falters

The Napoleonic Wars had been a euphoric time for Britain's financial sector, in comparison to which victory and peace were somewhat disappointing. The frantic search for high, rapid returns soon led to wild speculation – including large investments in a wholly imaginary Latin American country called Poyais, invented by Gregor MacGregor, a Stirlingshire-born former major in the army of Portugal. In 1825, the unmasking of Poyais and other almost equally laughable investment schemes led inevitably to a panic, in which more than 60 banks failed in England alone. The Bank of England itself was only saved by a large loan from the Bank of France. Building development across the country was suspended and, as we have seen, many estates and streets in Edinburgh were left substantially incomplete until the 1850s or later. Whilst many mapmakers including Ainslie [1804] and Wood [1830] depicted streets and houses that were in fact stalled at the planning stage, this map injects a welcome dose of reality, focusing on what actually existed on the ground. This accounts for its partly shaded buildings extending along streets shown only in outline.

Part of the western New Town, the Easter Coates estate west of Queensferry Road, was largely owned by William Walker, and laid out for development through a plan by Robert Brown in 1808. Building began in 1813 on Coates Crescent, and moved steadily north and east. Melville Street was the centrepiece of the scheme, and although building there was largely complete by the 1830s, the less popular sites at the corners of streets were not developed at this time. In fact, these gap sites would not be filled until the 1850s and 1860s, while streets laid out to the north of Drumsheugh Gardens were not built until the 1870s.

Shandwick Place was developed from 1806 by John Cockburn Ross, of Shandwick in Easter Ross, who owned land on the north side of the old road to Glasgow and the west. It

---

James Kay, *Kay's Plan of Edinburgh Containing all the Improvements to the Present Time* (1836)

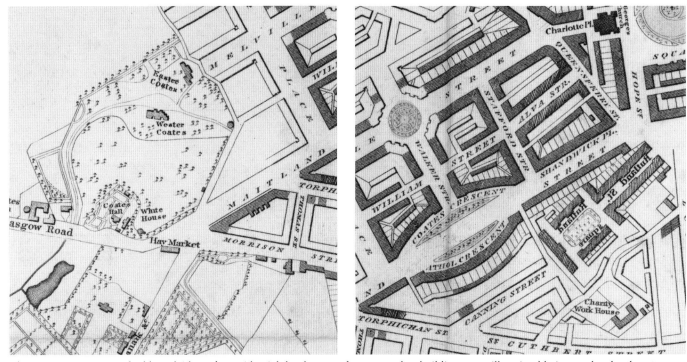

The Easter Coates estate had been laid out for residential development from 1813, but building was still noticeably incomplete by the 1830s.

joined up with what became Maitland Street, named after Sir Alexander Charles Maitland of Cliftonshall, who developed it from 1807. Here again, buildings peter out west of Manor Place, another area that lay empty until the 1860s. However, Rutland Square was acquired by Sir John Learmonth in 1825 and developed rapidly from 1830, as shown here.

James Kay was merely the publisher for this map; the real credit for its surveying, engraving and colouring rests with Kirkwood & Son. Robert and James Kirkwood were of course able to base it upon their existing, finely detailed map of Edinburgh [1817], and a note beneath the title confirms this provenance. An unusual error, however, can be seen in the text for Rutland Street and Square which appears in reverse.

Kay was a bookseller, based at the top of Leith Walk from the 1820s to the 1840s. He struggled financially, and was declared bankrupt in December 1832. His relationship, if any, to the brilliant Edinburgh caricaturist, etcher and sometime barber John Kay (1742–1826) is unknown. The map was originally published in 1829, and the darker and slightly bloated style of the numerals '36' in the date provide a clue to this, while in the lower margin on the original plate, Kay's earlier address of Blenheim Place has been hammered out. It is possible that, in prominently titling the map 'Kay's Plan', the bookseller (or the Kirkwoods) meant to capitalise on the fame of *John* Kay, all of whose ten children had pre-deceased him. In 1836, James Kay was selling off his book stock from his home in Haddington Place, and this map may have been re-issued to help business at a particularly difficult time.

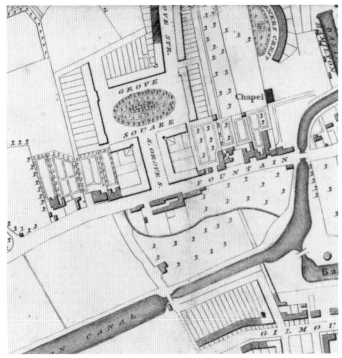
Grove Square at Fountainbridge, proposed but not implemented.

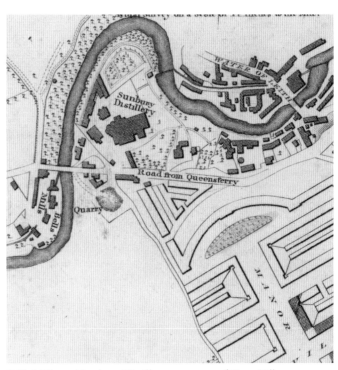
Bell's Mills and Sunbury Distillery, upstream of Dean Village. The projected streets shown here were only built at the end of the nineteenth century, and along less regular lines.

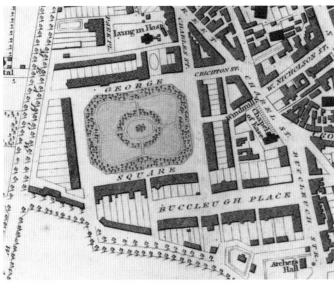
George Square environs, including the former maternity or 'Lying in' Hospital that occupied this site from 1793 to 1846.

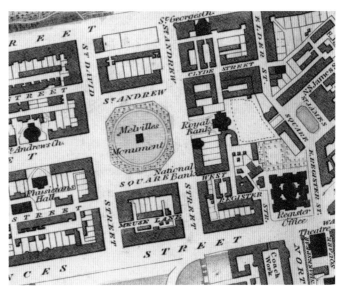
St Andrew's Square environs, with the new Melville Monument constructed in the early 1820s.

1843

Holyrood Park: a new pleasure ground

This map shows Holyrood Park at a time of transition in ownership, and into use as the public park we recognise today. Historically, from the foundation of Holyrood Abbey in 1128, the ownership of the park was shared between the royal family and the Abbey of Kelso, with St Anthony's Chapel attached to the Abbey of Kelso, and the Augustinian monks of Holyrood growing crops and operating a mill in the Hunter's Bog area. Following the construction of the original (northwest) block of Holyrood Palace in the early sixteenth century, James V had the park enclosed in the 1540s with a 'park dike', and in 1554 it was described as 'circulit about Arthus Sett, Salisborie and Duddingston craggis' [c.1610]. In the following centuries, the royal hunting park was also used as a sanctuary for criminals, and latterly debtors, who were only able to leave the park on Sundays. It was also used for military gatherings, including by Bonnie Prince Charlie's troops before the Battle of Prestonpans in 1745 (noted at lower right), and leased to the earls of Haddington as Hereditary Keepers, a privilege which they increasingly exploited commercially as time went by [1804].

The revolt of the Seaforth Regiment, noted on the eastern slope of Arthur's Seat, refers to a 1778 incident in which the 78th Regiment of Foot, having been marched to Leith to embark for the Channel Islands, came to believe that they had been sold to the East India Company. They mutinied and dug in on the hill in full posture of war, and probably could have remained there indefinitely, since the ordinary citizens of Edinburgh also believed the India/slavery tale and liberally supplied the rebels with provisions. One regular and two volunteer regiments surrounded their position, but negotiations led to final victory for the mutineers: 'a pardon to all of their number for all past offences; that all levy money and

---

William Nixon, *Plan of Holyrood Park Shewing the Proposed New Lines of Road* (1843)

arrears due to them should be paid before embarkation; and that they should not be sent to the East Indies'. The event has gone down in the city's history as the 'Affair of the Wild Macraes'.

By the early nineteenth century, thousands of tons of rock – both sandstone and harder freestone for house-building and road construction – were being quarried in Holyrood Park every year. The large quarry to the northwest of Hunter's Bog is clearly visible on nineteenth-century maps and on the ground today, but there were also several smaller quarries on the southwest escarpment of Salisbury Crags, visible from the whole South Side; Edinburgh residents vigorously objected to this rapacious exploitation of what was increasingly regarded as an important scenic amenity. Eventually, after Lord Haddington was awarded some £39,674 (well over £3 million at today's prices) for the loss of his interest in the park in 1843, the quarries were closed and the 580 acres of parkland were officially transferred to the Board of Woods and Forests.

In preparation for Holyrood's new era of public access, the Board actively worked with the Town Council and royal officials in commissioning William Nixon, the Clerk of Works in Edinburgh (1840–48) to undertake a survey of the park, and draw up plans for laying out carriage drives and footpaths. Had he done this a decade later, Nixon would have been able to use the Ordnance Survey Town Plan [1852], but at this time was forced to fall back on Kirkwood's excellent but rather dated plan of 1817 as a base [1817]. The agenda was clearly to redefine the park as a pleasure ground, particularly for carriage drives, exploiting its scenic and picturesque qualities. The large quarry to the northwest of Hunter's Bog was attractively filled with fluffy woodland, which also encircled the park itself. Bogs were to be drained; wild and precipitous slopes rendered accessible by smooth, curved carriage drives; new artificial lochs created (Dunsapie and St Margaret's); and the pestilential open sewer and 'irrigated meadows' to the north covered over and removed from view.

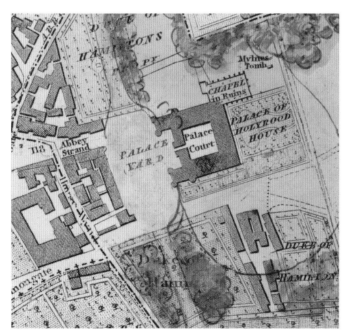

The proposed new avenues to the north and south through the former property of the Duke of Hamilton created a direct route to Holyrood Palace Yard.

The attractive fringe of woodland shown here by Dumbiedykes was proposed at this time, but would not be planted as a continuous strip for at least a century.

Apart from these new projected improvements, the only topographical update to Kirkwood's map was the inclusion of the Edinburgh & Dalkeith or so-called 'Innocent' Railway, running between Duddingston and St Leonards, which carried passengers from 1834.

Active work began in March 1843, making use of the large ranks of the destitute and unemployed to get the work done cheaply. By April it was reported that 200 men were actively working on the carriage drives, their numbers soon swelled by a further 'band of poor men' who had been working on draining the Meadows. As reported in the *Scotsman* newspaper, they were encouraged by the 'numerous groups of well-dressed persons, constantly visiting the works . . . communicating an air of gaiety and stir to the whole locality'. Although the initial drives were laid out quite quickly, approach routes and amenities were only constructed in the following decades, and did not quite follow the proposed lines illustrated here. The initial plans also included a thatched, rustic restaurant at the new Dunsapie Loch, but this idea was abandoned following energetic protests that referred to it as 'an unseemly howf'.

Whilst clearly intended for the gentry, Holyrood Park in practice saw a great range of other uses. Following a long campaign, the Town Council formally allowed two designated areas of the park to be used as bleaching grounds in 1849, although their locations near Hunter's Bog and Duddingston were hardly convenient. The park's long association with military gatherings and presentations also continued: by the 1890s, there were two large rifle ranges across Hunter's Bog, one of 1,800 feet, which were actively used until the mid twentieth century. The park was also a location for several large political gatherings and miners' rallies. During the Second World War some fringes of the park were developed as allotments and there was an anti-aircraft station with barrage balloons on the summit of Arthur's Seat.

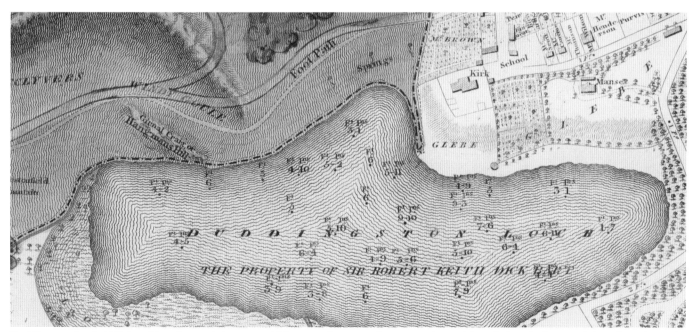

Duddingston Loch, Edinburgh's only remaining natural lake, was the scene of many meetings of the Edinburgh Skating Club in the eighteenth century. It is depicted in the background of the iconic Enlightenment-era painting *The Skating Minister*, widely if controversially attributed to Sir Henry Raeburn.

1847

Foul burns and fertile meadows

Over the centuries, Edinburgh dealt with its sewage in a piecemeal, localised manner, but the massive expansion of the city in the nineteenth century resulted in serious problems. Edinburgh's population quadrupled from 10,000 to 40,000 between 1560 and 1700, then more than doubled over the course of the eighteenth century. This would have been dramatic enough, but amid industrialisation and the rapid residential development of the New Town, South Side and West End, it nearly doubled again by 1841, to 165,000. Outbreaks of cholera in 1832 and 1848, and of typhus in 1847, were only the most visible results of the problem, yet for decades the Town Council dragged its feet over possible solutions.

Natural drainage into the Water of Leith was the traditional practical solution for western and northern areas, whilst the Canongate, Holyrood, Restalrig and eastern parts drained into the aptly named Foul Burn, which irrigated 250 acres of pasture land (worth £24 to £30 per acre per year in the 1870s) on the edge of Craigentinny and Lochend. Although the city's Medical Officers of Health certified that milk produced by animals grazed on this land was 'as safe as any other', the Craigentinny irrigation meadows also had their critics. As Francis Groome wrote in 1871,

> [t]his irrigation produced indeed large crops of herbage, but is a serious nuisance, loathsome to look upon, horrible to the olfactory nerves, and probably, even when the noxious gases arising from it are diluted with the pure air of the surrounding high grounds, not unaccompanied with material injury to the public health.

Five years before this map was published, Queen Victoria had elected to stay with the Duke of Buccleuch at Dalkeith rather than endure the stench in Holyrood Palace.

George Buchanan, *Plan of the Estate of Craigentinny* (1847)

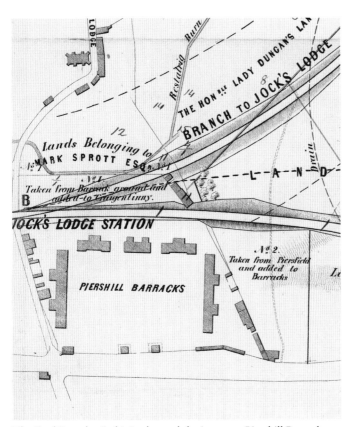

The Foul Burn by Jock's Lodge and the immense Piershill Barracks.

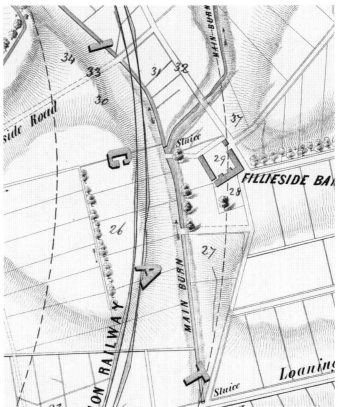

The Foul Burn running north by the farm of Fillieside Bank (today Craigentinny Golf Course).

This map proudly shows the 'Irrigated Meadows' in all their glory, from the principal source near Jock's Lodge through sluices and feeder channels fanning neatly outwards to the east. However, the sewage system is in fact only a backdrop to the main purpose of the plan: to promote a new Leith Branch Junction Railway from Jock's Lodge Station curving towards Seafield and Leith. The coastal route of the Leith Branch Railway itself (1846) is shown too, along with the North British Railway curving in from Portobello Station (also opened 1846) to Jock's Lodge. As is typical with such railway plans, the main proposed lines and 300-foot deviations are clearly demarcated on the ground, while the various landowners who would be paid compensation are also indicated: principally, in this case, the Earl of Moray and the Marquis of Abercorn. In the event, the Leith Branch Junction Railway was never constructed. As for the sewage, in spite of significant investments in new sewers in St Leonards and the Water of Leith in the 1870s and 1880s, it was not until the 1920s that the Foul Burn was covered over, and the Meadows sold by Lord Moray, to become housing and the Craigentinny golf course. The present clubhouse is on the exact site of Fillyside farm, shown towards the top of the map, which survived into the 1950s.

The large Piershill Cavalry Barracks was constructed in 1793, at a time when French invasion was possible but unlikely, whereas domestic radical subversion was considered an imminent threat, and the role of cavalry in domestic crowd and riot control was well established. The future novelist

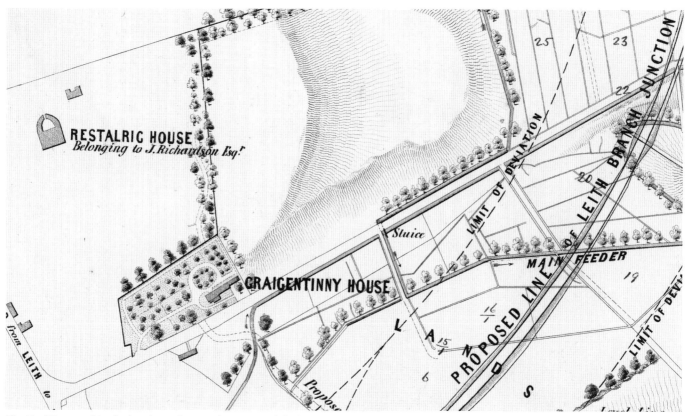

The Foul Burn and its feeder channels meandering gracefully by Craigentinny House.

Walter Scott joined the Edinburgh volunteer light dragoon regiment established in 1797, amid a positive flood of middle-class militarism deftly parodied by John Kay. In the event, working-class quiescence and the advent of civilian policing in Edinburgh in 1805–17 [1773] would make deployment of troops on the streets largely unnecessary; and it was Manchester, not Edinburgh, that became the notorious test case of sabres vs. unarmed protestors in 1819. Nevertheless, Piershill's importance only increased: field artillery troops were added, along with dozens of telephone lines by 1916, and it probably would have been the nerve centre of counter-invasion measures in the First World War had the Germans landed in Scotland. Cavalry horses from Piershill were regularly seen exercising on Portobello Beach until 1938, when the barracks was demolished to make room for the current council housing estate designed by E. J. MacRae (1881–1951).

The map was drawn by the surveyor and civil engineer George Buchanan (c.1790–1852), who was widely respected as an authority on public infrastructure projects. Following his report on salmon fishings on the South Esk near Montrose in 1827, he was often asked for advice on fishery disputes elsewhere. He advised on the construction of the Edinburgh Gas Light Company's chimney, 342 feet high, at the base of Calton Hill in 1847, and supervised the remarkable Scotland Street railway tunnel under the New Town, described by Robert Louis Stevenson as 'of paramount impressiveness to a young mind', with its 'dark maw' and 'many ponderous edifices and thoroughfares above'.

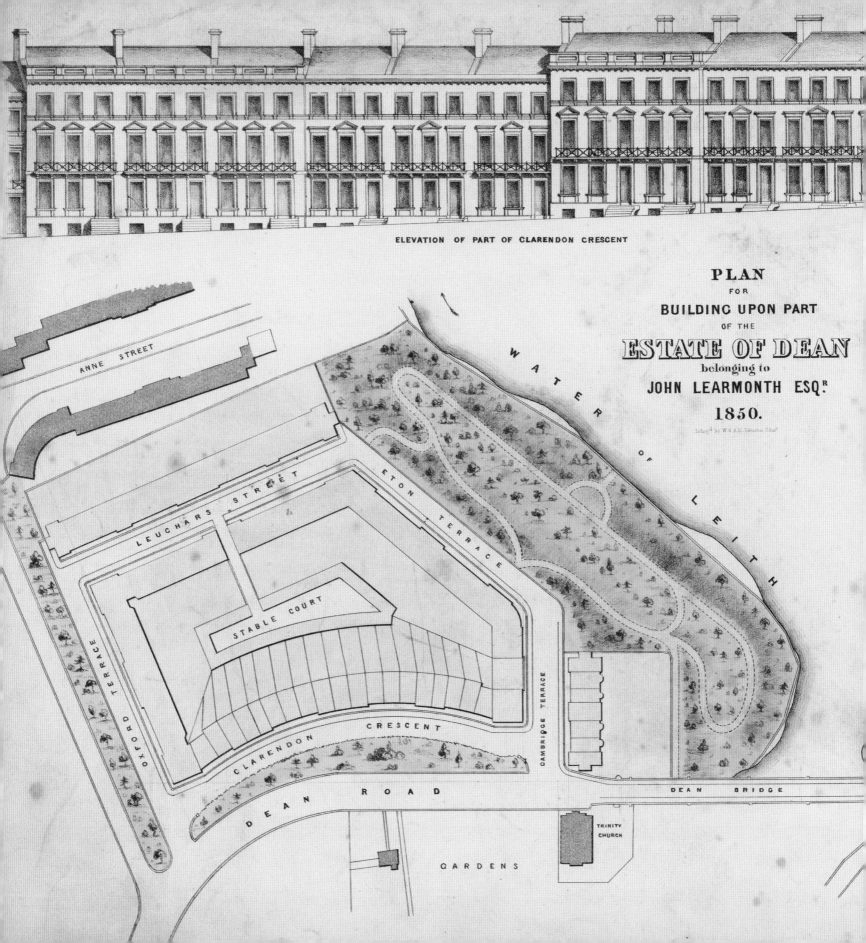

1850

Neo-classicism soldiers on: plans and elevations for the Dean Estate

This attractive building plan, along with its detailed architectural elevation, illustrates the precise long-term control that could be exercised over building through written feuing conditions and plans, despite wholesale changes to both the underlying landownership and prevailing architectural styles. The Dean estate lay to the northwest of the New Town on the north banks of the Water of Leith, and until the 1830s was primarily agricultural land. With the advance of the New Town and the increasing importance of a routeway through Dean Village to Queensferry, the area developed significant potential, which John Learmonth – who had made his money as a coachbuilder [1819a] and acquired the lands of Dean in 1826 from the Nisbet family – was keen to exploit. During his time as Lord Provost of Edinburgh in the early 1830s, Learmonth financed the construction of the lofty and majestic Dean Bridge (1831–32); largely designed by Thomas Telford, this carried the historic turnpike road by a new, direct course across the Water of Leith. It also opened up his Dean estate for feuing, hopefully to meet the Raeburn estate, which extended westwards from Stockbridge and had been laid out for building more than a decade earlier. Dreams of a quick return on his investment through a housing scheme similar to the Moray estate over the river remained unfulfilled, however, and a disappointed Learmonth sold much of the ground to the Heriot Trust.

That said, Learmonth's architect John Tait remained completely in charge of the building activity that took place from 1850 to 1853. (The OS large-scale town plan of 1852 shows buildings under construction as outlines.) Tait had studied W.H. Playfair's Regent Terrace, designed in 1825 and completed in 1833, and Tait's three-storey elevations – first-floor balconies with elegant ironwork, and windows topped

---

J. Tait / W. & A.K. Johnston, *Plan for Building on Part of the Estate of Dean belonging to John Learmonth, Esq* (1850)

with pediments, crowned with a balustrade above – lend his work a highly distinguished, yet fundamentally pre-1840 character. It should also be noted that the plan cleverly shortens the Dean Bridge to about a third of its real length, reducing the apparent distance to Randolph Cliff and Crescent.

Originally, Clarendon Crescent was to have been named for Queen Victoria, but the honour shifted to George Villiers, 4th Earl of Clarendon and Viceroy of Ireland (1847–52), whose repression of serious Famine-era unrest had made him temporarily very popular in mainland Britain, if lastingly detested in the island he ruled. The streets at either end were intended to be named after the great English universities of Oxford and Cambridge, but the houses on Cambridge Terrace were never built, and the name Eton Terrace was used instead from 1855 when buildings were constructed farther along. Leuchars Street, since renamed Lennox Street, was not developed for another decade (to revised plans by John Chesser), and never along its full length as depicted here. The delightful 'Stable Court' was also renamed, as Lennox Street Lane – though its profusion of painted timber garage doors announces its original purpose quite as effectively as any name could. John Learmonth's son Alexander, MP for Colchester almost continuously from 1850 to 1880, became a lieutenant-colonel in 1859, and inherited the Dean estate in the same year. He continued the trend of selling off parts of it, in his case to finance what became an increasingly expensive society lifestyle in London. In the following decade, Learmonth's bankruptcy allowed the remaining parts of the Dean estate to be acquired by the Edinburgh builder James Steel, who laid out new streets extending west towards Dean and north towards Comely Bank over the next two decades. Here, as elsewhere in the city, detailed plans and controls over building created an attractive skyline and a pleasing, if surprisingly dated, architectural consistency. Clarendon Crescent can still be seen largely as depicted here, an uncluttered sweep of terraces on a gentle incline and curve in the Queensferry Road, before the dramatic high-level entry to the West End across the Dean Bridge.

1850

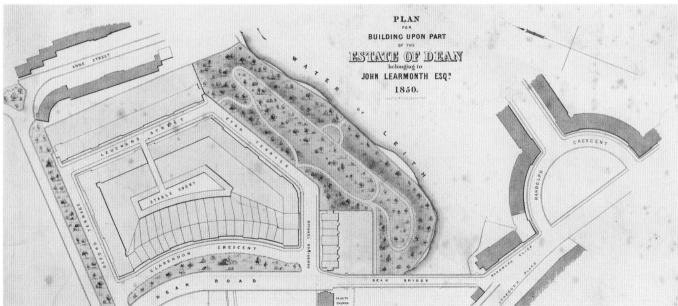

TOP. The elegant three-storey elevation of Clarendon Crescent, strikingly old-fashioned in light of its Victorian date, was designed by John Tait.

ABOVE. The plan cunningly shortens the Dean Bridge to about a third of its real length, making Randolph Cliff and Crescent in the New Town appear much closer to the Dean estate development than they are.

1851

The arrival of the railways

Although the Edinburgh–Dalkeith 'Innocent Railway' was constructed from 1826, primarily to reduce charges on coal from the Midlothian coalfields, it terminated at St Leonards, and it was only in the 1840s that railways (now pulled by steam locomotives rather than horses) reached the centre of Edinburgh. The first line to open ran from Scotland Street to Trinity from 1842, and – following its extensions in 1846 to Leith and Granton – it became known as the Edinburgh, Leith & Granton Railway. In 1847, it was extended south to reach Canal Street in the centre of Edinburgh by means of the Scotland Street tunnel under the New Town. This impressive feat of engineering, 1,000 yards long, has a gradient of 1 in 27, running only just below street level by Scotland Street, but falling to 49 feet below the ground at St Andrew Street. In the same year it was acquired by the North British Railway Company (NBR) who three years later extended the service by a train-carrying ferry from Granton to Burntisland [1834b] and on to Perth and Dundee. Its Canal Street terminus therefore appears here as the Edinburgh, Perth & Dundee Railway Station.

There had also been rapid progress in extending railways into Edinburgh from other directions. The Edinburgh & Glasgow Railway Company (E&GR) had opened a line from the west to Haymarket in 1842, and within four years had extended this line further east via tunnels beneath the Mound. Meanwhile, the NBR had extended their original line from Berwick-upon-Tweed to Haddington into Edinburgh from the east, tunnelling through the south flank of Calton Hill to a new station completed in 1846 on the former site of the Physick Garden [c.1690]. Following the merger of the E&GR and NBR in 1848, their station was often known as the 'Joint Railway Station', as shown here; note too the multiple small carriage/wagon turntables, with larger engine turntables just west of Waverley Bridge and above the Fruitmarket. It was

---

Alfred Lancefield / W. & A.K. Johnston, *Johnston's Plan of Edinburgh & Leith* (1851)

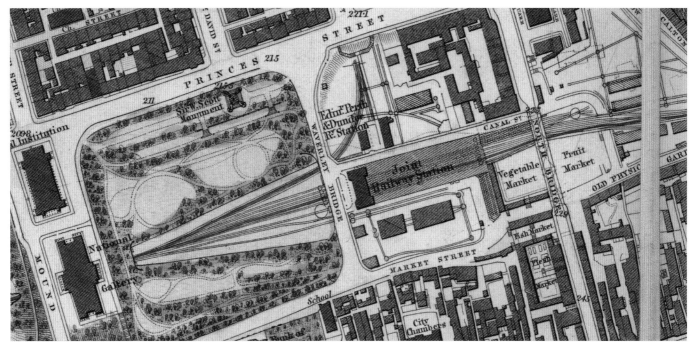

The Joint Railway Station housed early railway termini of the Edinburgh & Glasgow Railway, which entered it from the west, and the North British Railway, arriving from the east. The Edinburgh, Perth & Dundee Railway from the north entered through the remarkable Scotland Street tunnel under the New Town.

only following the opening of the NBR 'Waverley' route to Carlisle in 1862, and the reorganisation and expansion of the station during 1869–72, that it became more commonly known as 'Waverley'. In 1848, the Caledonian Railway opened an Edinburgh–Carlisle–London line from their Lothian Road station (on the far left of this detail); express trains completed the route in 12½ hours. The first public train from London to Edinburgh by the east-coast route arrived in 1850.

The new railway companies were hungry for space, always at a premium in the narrow valley. Perhaps the most serious architectural loss by this time was Trinity College Church, which was dismantled in 1848, following an agreement with the Town Council; only an apse in Chalmers Close off Jeffrey Street was ever built by way of replacement. Note also the new National Gallery on the Mound, designed by W. H. Playfair and constructed between 1850 and 1859, i.e. still eight years from completion at the time this map was published,

and just below the castle, the newly named Johnston Terrace, commemorating Lord Provost William Johnston.

When Johnston became Lord Provost, for three years from 1848, W. & A.K. Johnston briefly turned their attention to detailed mapping of the city, particularly utilising Alfred Lancefield as a surveyor. Lancefield worked as teacher of drawing and surveying at Merchiston Castle School, as well as an independent teacher of fortification and of military and civil engineering. He lived in Buccleuch Place in 1848 and South Castle Street in 1853. *Johnston's plan of the City of Edinburgh* (1850), a highly detailed sheet at a very large scale of 60 inches to the mile (1:1,056), was surveyed and drawn by Lancefield. It covered just the First New Town, and was intended to form part of a series, but in fact was the only sheet published. The following year, W. & A.K. Johnston issued this smaller-scale map, *Johnston's plan of Edinburgh & Leith in 1851*, covering the entire city and suburbs and measuring

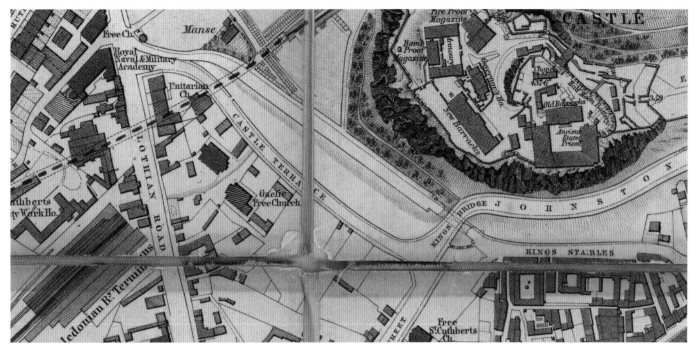

The initial terminus of the Caledonian Railway on Lothian Road. Johnston Terrace was named after the map's co-publisher, Lord Provost William Johnston.

some 52 inches wide by 66 inches high at a scale of 13 inches to the mile (c.1:4,800). It seems likely that the publication of the OS large-scale town plans, beginning in 1853, curtailed Johnston's large-scale series; but Lancefield successfully used both the 13-inch and 60-inch maps to advertise his skills, and was employed by the Police Commission working on drains during the 1850s. A revised and updated version of this plan was published in 1860.

The firm owned by William and Alexander Keith Johnston became one of Edinburgh's leading geographical publishers and cartographers with widespread international influence. Both men were educated at the Royal High School and trained as engravers with Lizars and Kirkwood respectively, before going into partnership in 1826. Keith Johnston also travelled in Germany and Austria, and through his collaboration with Heinrich Berghaus, Professor of Geography at Berlin, he went on to publish *The National Atlas* (1843) followed by *The Physical Atlas* (1848), both of which included innovative and influential thematic mapping. In the same year that they published Lancefield's map, Johnston produced the first physical globe ever made: 30 inches in diameter, it showed the geology, meteorology and hydrography of the Earth. A decade later, they brought out their monumental *Royal Atlas of Modern Geography*, a standard general folio atlas which would be revised many times in the years that followed. From the early 1860s W. & A.K. Johnston also successfully used lithography in mapping, reducing the cost of hand-colouring, and expanded their non-cartographic printing business. In 1837 they had moved their main office to St Andrew Square from the High Street, and in 1879 they opened their vast new EDINA works off Easter Road, with 50,000 square feet of floor space. This building survives today, having outlived the firm itself, which merged with G.W. Bacon of London in the 1940s, who in turn were swallowed up by Macmillan.

1852
1877

The Ordnance Survey's most detailed mapping ever

The Ordnance Survey large-scale town plan of Edinburgh, surveyed between 1849 and 1853, was the most detailed ever comprehensive map of Edinburgh and, arguably, the high point in the OS's mapping of the city. At 5 feet to the mile or 1:1,056, it was ten times larger than 6 inch to the mile mapping, the *de facto* standard for all rural areas, allowing a wealth of detailed information to be shown. The whole city was covered on 54 double-elephant sized sheets (each with map extents of 36 inches × 24 inches); if laid out collectively, these would cover an area 18 feet wide and 24 feet high. Especially when combined with the revised and expanded 66-sheet version of the maps issued in 1876–77, covering a somewhat wider area, they are an unrivalled resource for tracking change on the ground over a period of 25 years. In the twentieth century, the most detailed scale at which the OS mapped Edinburgh was 1:1,250, and economies had forced the exclusion of many details that had been shown in the Victorian era.

The Ordnance Survey's Edinburgh town plans were part

of a country-wide initiative, encouraged particularly by concerns over sanitation and urban health. The serious UK-wide cholera epidemic of 1831–33, with further outbreaks in the 1840s and 1850s, exacerbated by other killer diseases such as typhus, encouraged bodies such as the Poor Law Commissioners to recommend that the OS should map towns in detail in association with their mapping of counties. John Snow's famous distribution map of the cholera outbreak in London in 1854 also confirmed the value of urban mapping for disease prevention. Between 1847 and 1895, as the OS worked their way northwards mapping the counties of Scotland, 60 other Scottish towns were also mapped at the most detailed scales in their entire history.

Given the importance of the maps for improving urban sanitation, many features relating to lighting, gas, water supply, sewerage and drainage are shown, including lamp-posts (L.P.), fire plugs (F.P.), hydrants and water taps, while regular spot-heights and occasional bench-marks in feet above sea level are shown along all the roads. The maps show the divisions between all buildings, including tenements; the division between pavements and roads; and details of the wynds and vennels that are usually only numbered (at best) on other maps, even including flights of steps, and dotted crosses to indicate passageways. Businesses including pubs and hotels are named, and the internal layout of public buildings such as the Flesh Market, Poultry Market, City Chambers and St Giles are clearly shown. Unlike in many commercial maps, all the features were present on the ground, not merely projected or anticipated.

This detail shows the area around Cockburn Street before and after its construction in 1859–64. Originally named Lord Cockburn Street, it was largely encouraged by the Railway Station Access Company, set up in 1853 to improve communication from the Old Town to the three new stations that would merge to become Waverley. Henry Cockburn (1779–1854) was a Court of Session judge who took a keen interest in conservation and the preservation of Edinburgh's early architecture and character, and was vociferous in his criticism of insensitive modernisation and improvement. It is hoped he would have approved of the new street named in his honour; in any case, its gentle S curves, tidy baronial revival architecture and limited impact on views of the Old Town and Royal Mile were well received by others. By 1875, the Cockburn Association was founded as a formal body for the promotion of the values espoused by Lord Cockburn, and it continues to play an active part in conservation and planning today.

It is also worth noting the fundamental changes to Waverley station between the dates of these two maps, mostly undertaken between 1869 and 1873. Waverley Bridge was completely demolished and rebuilt as a wider structure, along the lines of Westminster Bridge in London, and narrow Canal Street replaced by a much grander street running down to the new booking offices. The platforms were also extended in number, width and length, running from just west of Waverley Bridge right along to Leith Wynd. At the time it was boasted that the station could 'permit twelve trains, without more than ordinary bustle or confusion, simultaneously to discharge or take in passengers'. The extension of the station meant other things had to go: the old Fruit and Vegetable Market just above the Flesh Market was moved in 1869 to the site of the former station built in 1847 by the Edinburgh, Perth & Dundee Railway [1851].

Overleaf left. Ordnance Survey,
Five Foot to the Mile – Edinburgh. Sheet 35 (1852)

Overleaf right. Ordnance Survey,
Five Foot to the Mile – Edinburgh. Sheet 35 (1877)

1852

The tourist city comes of age

This map, without the surrounding vignettes, was originally published in Edinburgh by Oliver & Boyd in their *Scottish Tourist* of 1852. This detailed and lengthy tourist guide to Scotland had first been published in 1825, to the general approval of Sir Walter Scott, whose letter of endorsement was printed in subsequent editions, each of which enlarged and corrected the original text. By 1852 the book was in its nineteenth edition, comprising 664 pages with 71 views and 17 maps, all engraved on steel by William Home Lizars (1788–1859). Nearly one-sixth of the work was devoted to Edinburgh and environs, describing the sights picked out on Lizars's map, with the remaining text describing 18 excursions around Scotland for the tourist, with 'the latest Railway and Steam-boat tours . . . minutely described'. The map continued to be included in later editions of the *Scottish Tourist*, as well as other guidebooks to Edinburgh, as late as the 1890s.

Lizars's carefully and charmingly engraved vignettes of Edinburgh landmarks convey the curious equivalency that developed, early in the railway age, between superficially similar tourist attractions of widely varying provenance. Heriot's Hospital, from the reign of Charles I, stands cheek-by-jowl with Donaldson's Hospital, a gigantic retro-Elizabethan structure completed just one year before the map itself. The neo-Gothic buttresses of the Walter Scott Monument (1840–46) are likewise twinned with the actually medieval Gothic lantern spire of St Giles. At top centre, perhaps most tellingly, a lady with a parasol has the option of surveying either the genuine medieval battlements of Edinburgh Castle to her right, or the mock-medieval ones of the city's Georgian prison to her left. The Grand Tour of the preceding century had taken the works of the Ancient Greeks and Romans as its theme and guiding principle. Now, 'the sights' of whatever period – even or perhaps especially the brand-new pseudo-Athenian temples built as banks, schools or the headquarters

---

W.H. Lizars, *Plan of Edinburgh* (1852)

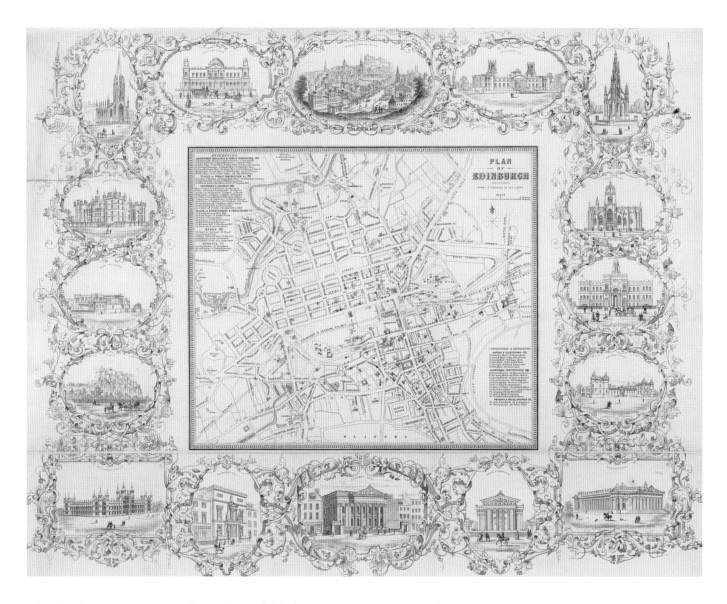

of medical associations – are all equally available for touristic consumption. The map proper continues this trend: the differing tints of its early colour-coding scheme are all that separate 'Places of Instruction' from 'Places of Entertainment', 'Churches and Chapels' from 'Banks' and 'Markets', 'Hotels' from 'Prisons'. Given the centrality of tourism to Edinburgh's present-day economy and self-image, this prescient map should probably be much better known than it is.

Based from 1817 onwards in St James Square in Edinburgh, Lizars was a talented artist and engraver who also drafted and published some of the most commercially successful Scottish maps and atlases of the early nineteenth century. He trained with his father, Daniel Lizars, who in turn had been apprenticed to Andrew Bell, who in his turn had learned his skills from Richard Cooper (1701–64), thereby not only continuing a long and successful lineage of engraving

Edinburgh through the tourist gaze: viewing the castle from Calton Hill.

Register House at the east end of Princes Street.

The Royal Scottish Academy on the Mound.

The Royal High School, Calton Hill.

expertise, but also contributing to late-Georgian Edinburgh's flowering as a centre for engraving. William showed early promise as a painter, exhibiting at the Royal Academy in 1812, but was forced to abandon that career owing to the death of his father in that year; he worked in partnership with his younger brother Daniel until 1819. He drafted and engraved numerous maps of canals, roads, harbours, towns and counties – particularly in Scotland – for many clients including the Edinburgh atlas publishers John Thomson [1822a] and William Blackwood; the Stevenson dynasty of civil engineers; and the Post Office, for use in their directories of Edinburgh; as well as for his own publications, including the *Edinburgh Geographical Atlas* (c.1840). He left a substantial estate of £8,219 5s 3d on his death, and his business was taken over by W. & A.K. Johnston.

1866

Clearing the Old Town slums

This map by the Edinburgh City Architects shows the proposed demolition of buildings in the Old Town to create new streets and improve sanitary conditions, as required by the Edinburgh City Improvement Act of 1866. It also illustrates the initial impact of Edinburgh's new Medical Officer of Health, Henry Littlejohn (1826–1914), and in particular his principle of 'piercing the closes with an airy street, [to] let in light, ventilation, and also the scavenger'. However, Littlejohn's relatively conservative aim of opening up specific overcrowded slums was developed into a much broader urban and architectural reconstruction project, particularly owing to the efforts of Lord Provost William Chambers and the City Architects, David Cousin and John Lessels.

With successive waves of immigration, particularly from Ireland and the Scottish Highlands, Edinburgh's population had nearly doubled in a generation, from 68,000 in 1801 to 136,000 in 1831. However, the financial slump that began in the 1820s [1836], resulting in a quarter-century dearth of building activity on all economic levels, exacerbated profound social and residential inequalities. Friedrich Engels, following a visit to Edinburgh in 1844, contrasted the 'brilliant aristocratic quarter' with the 'foul wretchedness of the poor in the Old Town'. The Scottish system of feuing land, where feu-duties were set at a fixed level for all time, encouraged their 'front-loading' to compensate for the inevitable erosion of real value, but this could be compensated for via high-density housing, and by building upwards. Many of the Old Town tenements were six or even eight storeys high with no running water or sewerage, and in a ruinous state. In November 1861, a tenement in the High Street collapsed 'with a hideous uproar' killing 35 inhabitants. A carving on the nearby Bailie Fyfe's Close – 'Heave awa' chaps, I'm no deid yet!' – records the cry of a young boy trapped in the rubble.

Littlejohn occupied a number of posts in public and

David Cousin and John Lessels, *Revised Plan of Projected Improvements of the Old Town of Edinburgh* (1866)

forensic medicine, including that of Police Surgeon for the Edinburgh Town Council from 1854, and became Edinburgh's first Medical Officer of Health in 1862 under the terms of the General Police Act (Scotland). At the age of 29, he had performed an impressive and original statistical analysis of the causes of death and morbidity in Edinburgh, underpinned by a coherent new set of Sanitary Districts, and this provided the crucial empirical basis of his famous *Report on the Sanitary Condition of Edinburgh*. Broad-ranging but also practical and sensitive, its 120 pages of text, tables, maps and diagrams covered housing and population densities, fresh water and sewerage, industrial pollutants, cow byres, bakeries, burials and cemeteries.

The *Report* was well-received by many, particularly the publisher William Chambers (1800–1883), an enthusiast for sanitary reform who was elected Lord Provost in November 1865. Chambers immediately undertook a three-month detailed perambulation of the Old Town with David Cousin, the City Architect. The following year, Cousin and fellow architect John Lessels drafted their important series of maps proposing improvements, which were made available for public consultation after just three months. Whilst Chambers remained vociferous in his support for the improvements, there was widespread criticism: some felt that the scheme's geographic focus was misdirected, and others disliked how hastily it had been conceived; but there was particular concern over its high costs. The *Report* proposed to displace 3,257 families, or about 15,000 people, and even though one-third of the forecast cost of £300,000 would be offset by the sale of building stances, the remaining two-thirds would have to be funded by higher personal taxes over a period of decades. Even after much lobbying, the final scheme was only approved in the teeth of significant opposition.

This map is from the Edinburgh Improvement Act volume of November 1866, with maps drawn by W. & A.K. Johnston. This detail is from the initial small-scale map at the outset, though the volume includes more detailed plans and elevations of each of the numbered areas. On this map, new thoroughfares and the widening of existing ones are tinted blue; new open spaces (intended to be permanent) are yellow; and new blocks of buildings are red. By May 1868, work had begun on St Mary's Wynd, which was doubled in width by the removal of buildings on its east side; it has since been known as St Mary's Street. The transformation of Blackfriars Wynd into Blackfriars Street proceeded on nearly identical lines. The other major developments followed between 1872 and 1876, and included the new Jeffrey Street extending from St Mary's Wynd/Street and curving down to Market Street. It was named after Lord Jeffrey (1773–1850), an Edinburgh judge, advocate and editor of the *Edinburgh Review*. This continued the naming tradition established more than a decade earlier by nearby Cockburn Street [1852/1877], which it closely resembled both topographically and architecturally. The new Chambers Street, named after the Lord Provost and constructed at the same time, was perhaps the boldest stroke of all, demolishing much of the former Argyle and Brown Squares at its western end, and forming a new avenue by the new Industrial Museum and Old College. The main work was incomplete when Chambers's initial term as Lord Provost ended in 1869, so he agreed to run again and was re-elected. A statue of Chambers by John Rhind, unveiled in 1891, still stands in Chambers Street today.

The only proposed street shown in Littlejohn's *Report* ran mid-way between the High Street and Cowgate, and parallel to them between South Bridge and St John's Street, opening up closes in the Tron and Canongate slums, and this is partly reflected in numbers 3 and 4 on this plan. But these proposals, along with those numbered 2 and 5, were never implemented; nor were the blocks marked 'A A' just north of Old College, reserved for possible new college buildings, ever used for that purpose.

The sanitary/improvement works' final cost came to over a quarter of a million pounds – up to £322 million today – and removed 2,800 buildings, including some that were the last survivors of their type. However, as a result of this and other measures, the death rate by 1884 had fallen by as much as 43 per cent in the central Tron district, and by 20 per cent across the whole of Edinburgh: a remarkable vindication of the city's mid-Victorian health policy in both general and specific terms.

W. & A.K. Johnston used Lancefield's map [1851] for their topographic base, with only minor updates for the most important changes on the ground, and it thus includes some details that had no connection to sanitary improvements: for example, the fact that the signing of the Treaty of Union of 1707 took place (partly) in the Earl of Moray's 'Summer Ho[use]' in the South Back of Canongate. The Victorian era was arguably the first, and the last, moment at which the Treaty would have been generally regarded as a benign object of patriotic veneration. In fact, the Summer House had been used precisely because it was a 'secret spot': the lives and property of pro-Union politicians were not safe from howling anti-Union mobs in the winter of 1706–07, despite 'the whole army', including a regiment of horse guards, having been moved into the city specifically to protect them (according Robert Chambers's account from 1833). Jacobites and others continued to criticise the Treaty's validity or morality down to the death of the last Jacobite claimant, titular Henry IX, in 1807, and other voices joined them, including that of Robert Burns; while the modern form of organised nationalism would arise in Scotland not long after the First World War.

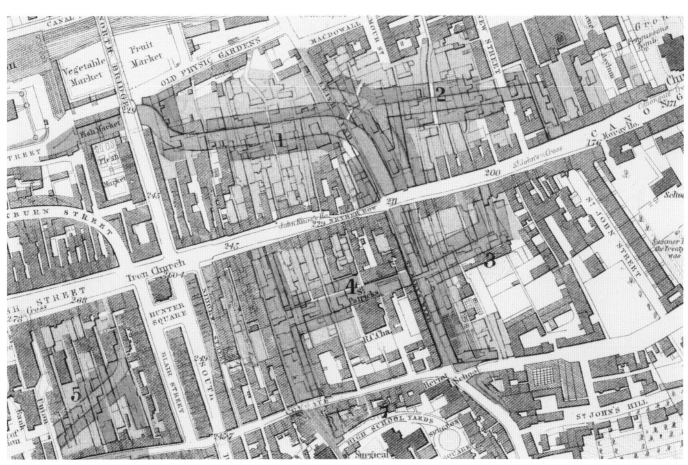

The central part of the Improvement proposals, creating Jeffrey Street, St Mary's Street, and Blackfriars Street. The proposed developments numbered 2, 3, and 5 were not implemented.

1869

The politics of drains and sewerage

In the second half of the nineteenth century, W. & A.K. Johnston produced a number of infrastructure plans for urban administration, often using their standard Post Office Directory street plan as a base map with different thematic overlays printed lithographically. This map from 1869, entitled 'Drainage', captures the city's fairly inadequate system of sewers and drains at a time of serious conflict over how to improve them – and, of course, who was to foot the bill.

The map gives something of a false impression of an integrated system or network; in reality all that existed was an *ad hoc* series of storm drains, designed only to reach the nearest watercourse by the quickest routes, and to which many tenements were not connected at all. To the extent that things did connect, there were effectively three sewers for the whole of the Old Town at this date. One of these took water off the Nor' Loch eastwards along the North Back of the Canongate; a second went by the Cowgate to Holyrood Park; and a third ran from the environs of George Square, also to Holyrood Park. These three sewers combined to form the Foul Burn near the Clock Mill, flowing through the notorious 'irrigation meadows' to the sea [1847]. The other districts of the city to the west and north drained principally into the Water of Leith, which at most times did not have a sufficiently powerful flow to carry sewage away. The problem was aggravated by increasing abstraction of water in the Pentlands for Edinburgh's water supply, and the intensive use of water by industrial operations along the Water of Leith. This led to regular complaints, particularly by New Town residents, about the disgusting state of the Water of Leith that flowed in such close proximity to their gardens and villas.

Following the global cholera epidemic of 1829–51, a conviction arose that polluted water was a carrier of disease, and this prompted efforts to improve the quality of the Water of Leith; but (as usual) there were serious arguments over the

W. & A.K. Johnston, *Drainage* (1869)

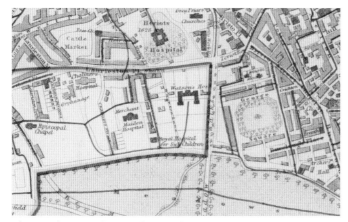

The area between the Meadows and Heriot's, with drains flowing west into the Water of Leith demarcated in yellow, and those flowing east towards Holyrood and the 'irrigation meadows' in pink.

expense associated with proper works, and who should pay. The mixed responsibilities for the different pollutants in the Water of Leith, industrial and domestic, and arguments between the Edinburgh and Leith Town Councils over the proposed solution of transporting Edinburgh's sewerage to Leith through a new sewer, caused heated disputes. Well-founded allegations of corruption were also made against the Water and Sewerage Commissioners in 1865. Given this tense political background, Henry Littlejohn had trod carefully indeed in his *Report* [1866], focusing more on poverty as a cause of morbidity and mortality than on sewage, and his research showed that – contrary to popular expectations – rates of diarrhoea and dysentery were not particularly high alongside the Water of Leith.

It was not until the Edinburgh and Leith Sewerage Act (1864) that parliamentary plans for modernisation were completed and approved. The 1864 interceptor sewer, into which many of the earlier nineteenth-century sewers were soon connected, was a considerable improvement, even though it extended only as far inland as Roseburn. The second, deeper sewer was constructed inland to Balerno in 1889, the same year in which a body of commissioners was set up to supervise the Water of Leith, but it was only in 1920 that these matters came under the direct control of local government.

Sewer improvements shifted the problem from the Water of Leith to just beyond the mouth of the (greatly expanded) Leith Harbour, where the raw sewage entered the Firth of Forth – a new problem that was only addressed with the construction of the Seafield sewage works in the 1970s. However, the abundance of certain types of duck in the Forth, such as the scaup, pochard and goldeneye, during the first three-quarters of the twentieth century has been directly attributed to the presence of substantial quantities of effluent, and sightings of these species have rapidly declined since the 1980s.

In addition to drainage matters, the map provides a very clear picture of the dramatic recent expansion of Leith Harbour, via the addition of the Victoria Dock (completed 1851), the Prince of Wales Graving Dock (1858) and the brand-new Albert Dock (1869), Scotland's first to use hydraulic cranes. The massive new Leith Gas Works at 1–5 Baltic Street was begun in 1835. It was also now possible to travel all the way from Edinburgh to Leith on city streets – albeit mostly still with open country on one side or the other.

The 'Edinburgh Artillery Militia Head Quarters' on the east side of Easter Road predates the French invasion scare of 1859, but is a reminder of the intense socioeconomic importance of the Volunteering movement at this time. In the event, not a shot was fired in anger between the French Second Empire and the British, yet hundreds of small groups of part-time soldiers were formed, often wearing fanciful uniforms of their own devising. They paid for their own rifles, and were deemed 'effective' if they received eight days of training over four months. Within two years, 211,000 British men and boys had volunteered, rising to a third of a million by 1908, when the movement (along with older part-time units including the Edinburgh Artillery Militia) was subsumed into the newly created Territorial Army. It has even been suggested by Professor Sir John Keegan that the sudden, widespread popularity of uniformed paramilitary 'rifle clubs' in the USA, modelled very closely on the British volunteers of 1859, was one of the immediate triggers of the American Civil War of 1861–65.

Drainage

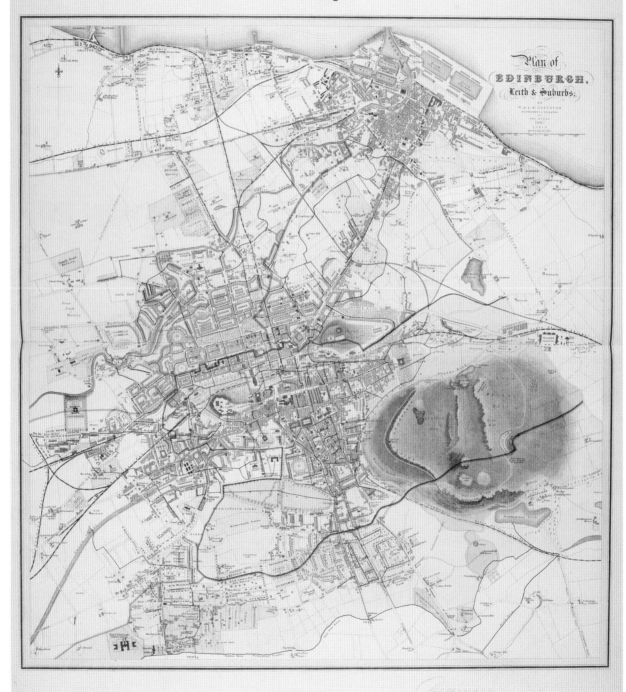

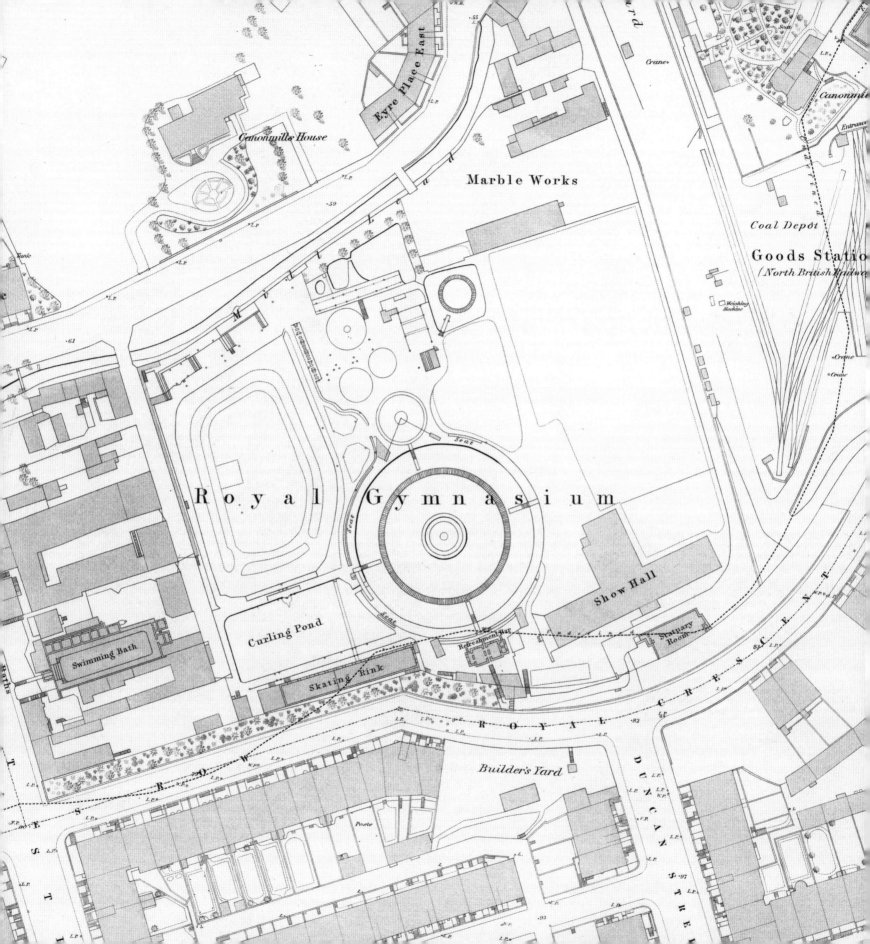

1876

Park of wonders: the Royal Patent Gymnasium

Royal Crescent, planned as part of the northern New Town, originally curved gently around mill lades flowing into Canonmills Loch to the north, which in wetter weather spread out over the low-lying meadow or Canonmills Haugh. This area – which today houses the George V Memorial Park and a car park for the Royal Bank of Scotland – was home from the 1860s to the quite exhilarating Victorian venue shown here.

The Royal Patent Gymnasium was the idea of John Cox of Gorgie House, a businessman and philanthropist who firmly believed in the benefits of the type of mass open-air recreation that had recently been pioneered in Germany, particularly by the 'father of gymnastics', Friedrich Jahn. With the opening of the new Scotland Street Railway Station in 1842, and the draining of Canonmills Loch five years later, the site had developed great potential, which Cox quickly exploited on an ambitious scale: packing it with innovative rides and fairground-style exercise equipment, and generally stretching the concept of 'physical education' (which was itself relatively novel) until it was transformed into something like the modern amusement park. It was opened in April 1865 in the presence of the magistrates, councillors and numerous worthy inhabitants of Edinburgh, and underwent various enlargements and improvements in the years that followed.

The enormous concentric circular structure was a 'patent velocipede': a paddle merry-go-round 160 feet in circumference that could accommodate 600 people. Clearly this doubled up as a 'patent rotary boat or "Great Sea Serpent", seated for 600 rowers' who could embark and disembark at four different piers. The structure was suspended by wires from a central post over shallow water and propelled around by the participants. At other times, the central pond would be filled with 'interesting varieties of small boats and canoes . . . propelled by various novel methods', as well as the 'Prince Alfred Wreck Escape', a simulated shipwreck experience that could be

---

Ordnance Survey, *Edinburgh – Five Foot to the Mile. Sheet 24* (1876)

enjoyed (if that is the word) even by visitors who did not know how to swim. The scale of the Ordnance Survey large-scale town plan of 60 inches to the mile allows excellent detail of permanent features to be shown; smaller-scale maps of Edinburgh capture at best the location of the Gymnasium only.

The 'Show Hall', exhibition hall, or athletic hall to the right included velocipedes, 'with the largest Training Velocipede Course in Scotland' and 'an instructor always in attendance'. In winter, the grounds could be easily flooded for skating and curling (lower left), and the rinks lit by hundreds of lights so that 'the scene, with its musical accessories, is one of wonderful brightness, gaiety, colour and incessant motion'.

Towards the top of the map are an array of swings, rides and other contraptions. The giant's see-saw, known as 'Chang', was a pendulum swing 100 feet long and 7 feet broad, accommodating up to 200 people and elevating them 50 feet into the air. There was also a self-adjusting trapeze, enabling gymnasts to swing by their hands a distance of 130 feet from one trapeze to the other; and a compound pendulum swing, holding 100 people, 'kept in motion by those upon it'. For those seeking greater thrills, the 'New Grand Planetarium Swing', introduced in 1877, represented the Earth, Neptune, Jupiter and Saturn, with each planet making a vertical orbit of nearly 300 feet, and accommodating up to 12 'inhabitants' at a time. Complementing these grand contraptions were more workaday accessories: 'Giant's strides, giant's leaps, horizontal and rotary ladders, parallel bars, vaulting poles, sloping-sliding ropeways, stilts, quoits, balls, etc, etc.'

With a cheap entry price of sixpence (threepence for children under 12), and half-price offers for groups, crowds flocked to the Gymnasium from the day it opened, and regular advertisements in the Edinburgh Post Office Directory in the 1860s and 1870s attest to its enduring success – not least in self-promotion. Regular special events and tournaments attracted competitors from all over Scotland and further afield, with large crowds of spectators. A military or other band on Saturday or holiday afternoons – weather permitting – added to the carnival atmosphere.

Of course, not everyone appreciated the fun, especially local residents; and when the Gymnasium applied to renew its licence in 1900 there was a vigorous campaign against it. As local resident George Matthewson recalled in his letter to the *Scotsman* on 31 May 1900: 'Besides the annoyance caused by the monotonous din of a steam organ, the discharge of firearms at shooting galleries, and the nightly visitation of a yelling mob . . . there was an entire absence of any sanitary arrangements and other evils.' By this time, in any case, the popularity of the Gymnasium had waned, no doubt because the novelty had worn off, but also because, after 1880, the latest thinking in physical education had begun to discountenance German-style gymnastics – with its emphasis on costly and complicated equipment – and instead to favour the 'light' or 'Swedish' form of gymnastics, now known as calisthenics, which required no equipment at all. Neither system of exercise, however, would ever again rank as participatory mass entertainment; and on the Ordnance Survey map of 1905 we find the site occupied by the stand of 'the Saints' – St Bernard's Football Club – who enjoyed their glory years in the early twentieth century.

THE
ROYAL PATENT GYMNASIUM,
THE NEW AND INCREASING WONDER OF EDINBURGH,
ROYAL CRESCENT PARK, adjoining Scotland Street Station.

THE PATENT ROTARY BOAT,
'GREAT SEA SERPENT,'
Is seated for 600 Rowers, embarking and disembarking Passengers at Four different Piers, at opposite sides of the Island, at the same time!!!

THE PATENT VELOCIPEDE
PADDLE MERRY-GO-ROUND
Is 160 feet in circumference, and can accommodate 600 Persons.

THE PATENT
Giant's Sea-Saw 'CHANG'
Is 100 feet long by 7 feet broad, and can accommodate 200 Persons, elevating them to a height of 50 feet.

THE PATENT
SELF-ADJUSTING TRAPEZE
Enables Gymnasts to Swing by the hands a distance of 130 feet from one Trapeze to the other.

THE PATENT
COMPOUND PENDULUM SWING
Holds about 100 persons, and is kept in motion by those upon it.

There are also interesting varieties of
SMALL BOATS AND CANOES ON THE PONDS
Propelled by various novel methods.

THE PRINCE ALFRED WRECK ESCAPE;
Also to be seen Specimens of
SWIMMING STOCKINGS, STILTS, & STOUPS.

The Giant's Strides, Giant's Leap, Horizontal and Rotary Ladders, Parallel Bars, Vaulting Poles, Sloping-Sliding Ropeways, Stilts, Quoits, Balls, etc. etc.

Admission, 6d.; Children under 12, 3d. Season Tickets at moderate prices.

Schools and Institutions, en masse, *Half-price.*

A MILITARY or other **BAND** on Saturday or Holiday Afternoons, Weather Permitting.

1879

Mapping the speed of sound for the Time Gun

This striking map aims to show the number of seconds after the Edinburgh One O'Clock Gun was fired that its sound would be heard across Edinburgh. Whereas the smoke of the gun could be seen immediately, those on Calton Hill would hear the sound about four seconds later, and those in Leith and Newhaven, a couple of miles away, would hear it 11 seconds later. As stated on another version of the map:

> For every additional circle of distance at which the observer may be from the Castle, he should subtract one second, due to the measured rate at which sound travels, from the instant at which he hears the REPORT of the Time-Gun in order to obtain the exact moment of the fire; or, One o' clock.

A time gun had been suggested by Edinburgh businessman John Hewitt, who had seen one in Paris in the 1840s, and a 5½-foot time-ball was constructed in 1852 on top of Lord Nelson's Monument on Calton Hill. The ball was raised to a height of 15 feet shortly before one o'clock, and made to drop exactly on the hour, as an aid to setting ships' chronometers for the calculation of longitude. This could be supplemented on foggy days by the report of a cannon, but there were concerns that a gun on Calton Hill would shatter windows in the vicinity; and it was only in 1861 that a gun was set up on Mills Mount Battery in the castle to synchronise with the ball. The main scientific credit for this goes to Charles Piazzi Smith (1819–1900), who was appointed Astronomer Royal for Scotland and Regius Professor of Astronomy at Edinburgh University in 1846. Smith took his middle name from the famous Sicilian astronomer Giuseppe Piazzi, and worked for his father's friend Thomas Maclear in the Royal Observatory at the Cape of Good Hope, where he had seen another time-ball in action. He had also presumably read accounts of the

---

W. & A.K. Johnston, *Plan of Edinburgh, Leith & Suburbs* [showing the Time Gun] (1879)

groundbreaking experiments undertaken at the turn of the seventeenth to eighteenth century by the Rev. William Derham of Upminster in Essex, who had calculated the speed of sound with reasonable accuracy using a 16-foot-long telescope, two synchronised watches and an assistant armed with a fowling-piece.

From 1861, Piazzi Smith's system operated via an electrical linkage between the Royal Observatory adjoining the Nelson Monument and the Time Gun in Edinburgh Castle. This consisted of a single-span, 3/4-mile-long electric cable passing over Waverley Valley, laid by a squad of sailors from Leith. In 1873, to reduce the strain placed on the Nelson Monument by the weight of this apparatus, the cable was re-laid via the Old Post Office, the New Post Office, St Giles and the steeple of the Tolbooth Church. Then, from 1896, the current was sent along the General Post Office cables to the new Royal Observatory that had been built in 1892 on Blackford Hill.

The map highlighted the impressive precision of the equipment used, with a helpful explanation of the differences between the speed of light, electricity and sound, stating that the clock 'shows always the true time, as close as it can be ascertained, by human means'.

The time-ball mechanism and clock were designed by Maudsley, Sons & Field (who had previously made a time-ball for Greenwich Observatory) and installed by Edinburgh clockmakers F.J. Ritchie & Son, who had premises on Leith Street at this time. The ball, still dropped every day, is the only

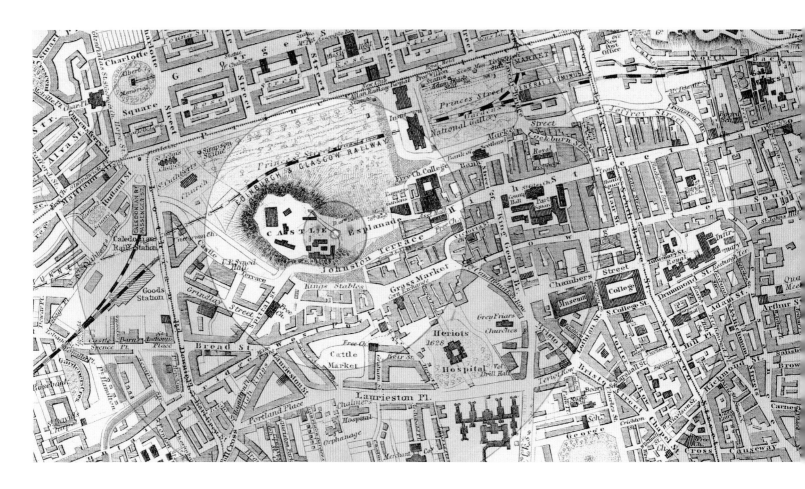

hand-operated one remaining in the world, and was recently subject to a full restoration.

In practice, the speed of sound was influenced by wind direction, air temperature and atmospheric pressure, and not simply the neat concentric circles shown here, and this may reflect Piazzi Smith's misconceptions. He had a mixed career, due to a lack of formal mathematical training, and was forced to resign from the Royal Society in the 1860s following his misinterpretation of data from the Great Pyramid. As an evangelical Christian, he shared the widely held belief that the British inch was associated with the sacred cubit of the Bible – part of a wider campaign at the time to resist the imposition of the French metric system in Britain – and that the pyramid embodied basic scientific information built under divine guidance by the ancient Israelites. His relations with the university became strained, and funding for the Royal Observatory grew increasingly scarce, although he continued his own research, from a new house on Royal Terrace, where he and his wife lived from 1871.

W. & A.K. Johnston used their standard engraved map of Edinburgh at 6 inches to the mile (1:10,560) as a base for this map and several variants, with different colours. In 1861 and 1862, for instance, it had appeared in the Edinburgh Post Office Directory with simple red concentric rings. It was also issued in 1881 as *Hislop's Time-Gun Map of Edinburgh*, with green rings, by the Edinburgh publisher Alexander Hislop of Elder Street.

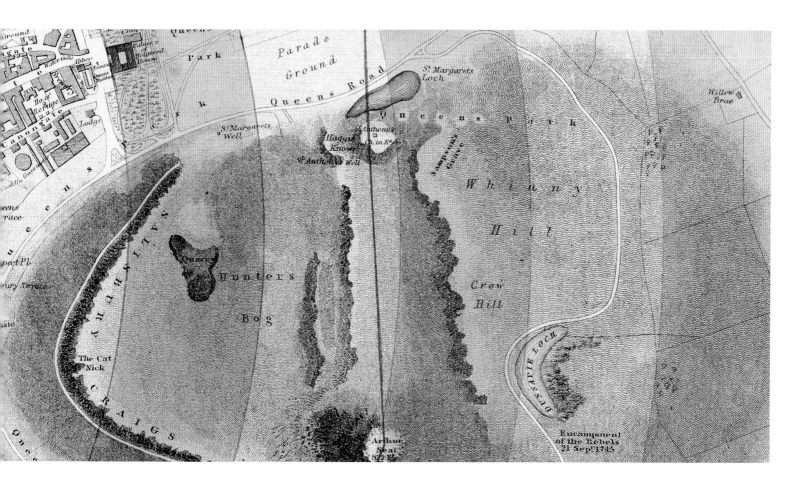

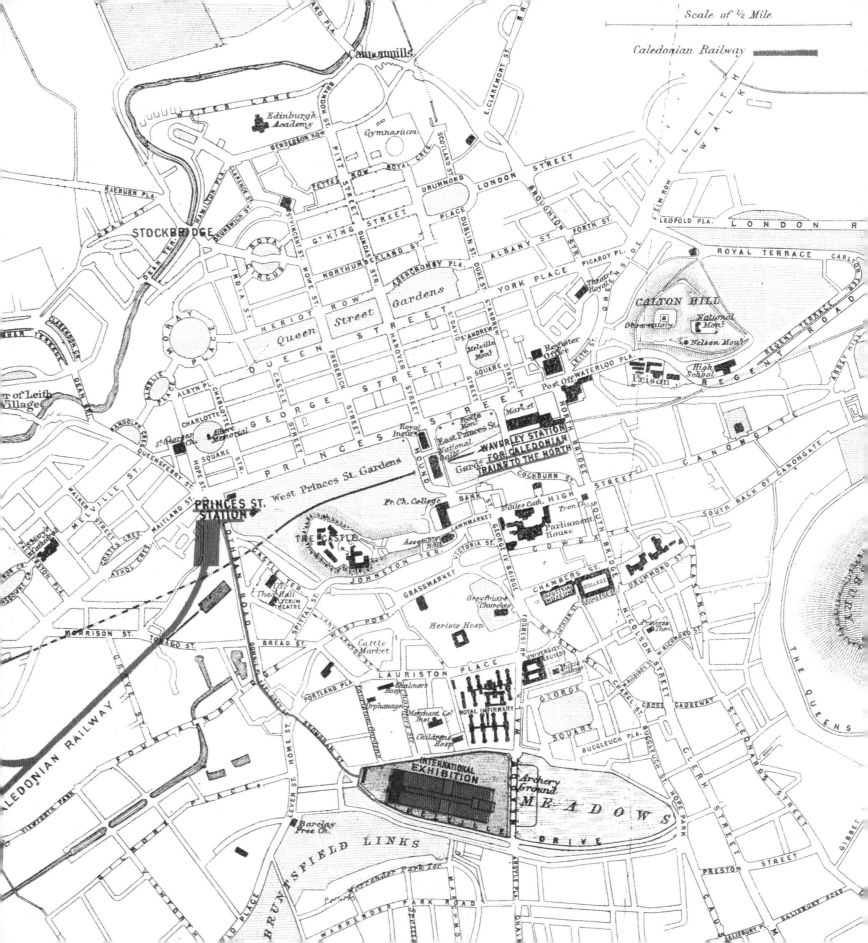

1886

A very Scottish international exhibition

Inspired by the successes of the Great Exhibition in London in 1851 and its follow-up in 1862, which both drew crowds of over six million, Edinburgh's International Exhibition of Industry, Science and Art ran from May to October 1886. All such events, including London's Great Exhibition, owed a great deal to the popularity of the Tenth French Industrial Exhibition of 1844, and while some, such as the Philadelphia Centennial Exhibition of 1876, are particularly well-remembered, nearly 40 were held in a dozen countries during the 1880s alone. Liverpool's International Exhibition (of Navigation, Commerce and Industry) opened in the same month as Edinburgh's. The trend would finally be derailed only by the Second World War, from which point any similar events would centre on peace and international co-operation rather than industry and trade. The 1886 Edinburgh exhibitors and visitors both came from far and wide to promote and celebrate the material progress of the age; but it was also intended to be a distinctively Scottish event.

The cover of this promotional leaflet and map (p. 198) provides a bird's-eye view of the exhibition building, with its impressive pavilion facade facing Brougham Place. This sported a 120-foot-high dome, decorated with the signs of the zodiac, and in a niche, a figure of Great Britain receiving the arts and industries. A statue of Minerva, goddess of wisdom and the liberal arts, perched beneath a large winged figure on the dome itself, representing Fame. Inside the main hall, which could hold up to 10,000 people at a time, corridors and galleries filled with more than 20,000 exhibits promoting the wonders of the Victorian age and Empire competed for visitors' attention.

J. Bartholomew, *Plan of Edinburgh* for the
Caledonian and North Western Railways' guide to
the Edinburgh International Exhibition (1886)

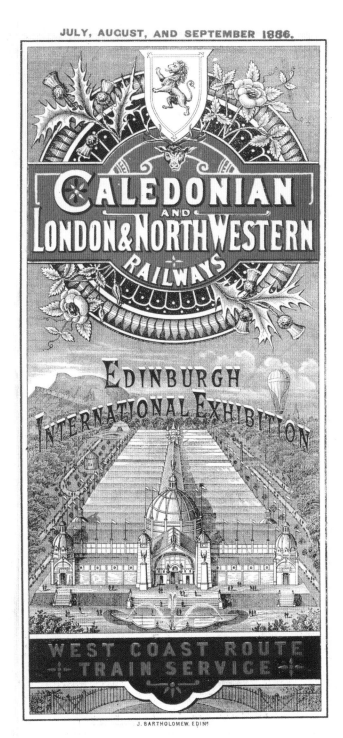

The exhibition boasted a 250-foot-long railway with four locomotives and a tender, as well as a tram ferrying visitors from Brougham Place to Middle Meadow Walk. There were educational appliances, Italian furniture and marble, a grand organ donated by the Bishop of London, violins from Prague, and exhibits related to mining, pottery, bread-making and sugar-refining, as well as printing and paper-making 'in which Edinburgh excels'. Successive contingents of girls brought in from Shetland and Fair Isle demonstrated carding, spinning, dyeing and knitting. Edinburgh firms promoted extravagant wares: Jenners offered a case of handmade outfits, including a silk dinner dress adorned with pearls and metal beads; Stuart & Co. unveiled their new 'granolithic' concrete, using it to produce two 22-foot-high Corinthian columns; and J. & G. Stewart sponsored a National Exhibit of Scotch Malt Whiskies. Two thousand paintings hung alongside a replica Grecian temple, its pillars and steps covered in state-of-the-art linoleum from Kirkcaldy. There were also model dwellings on show, fitted with the very latest gadgets including gas lamps, ventilating tubes and free-standing cast-iron baths.

Perhaps the major highlight of the exhibition, however, was 'Old Edinburgh': a large-scale reconstruction of a typical seventeenth-century Edinburgh street, featuring famous buildings of the Old Town that had long since been demolished. These included the Netherbow Port, the Black Turnpike [1793], the Old Tolbooth, the house of Major Weir (which was burnt in 1670 when its occupant was accused of satanic practices), the Royal Porch into Holyrood, the French Ambassador's house (taken down in 1829 during the construction of George IV Bridge), and Paul's Work (a charitable workhouse in Leith Wynd). Contemporary reports praised 'Old Edinburgh's' painstaking attention to detail and excellent imitation of medieval architecture and stonework, and the concept would be imitated on an even grander scale by the organisers of the Glasgow International Exhibition of 1901.

An 1827 Act of Parliament forbade permanent building on the Meadows, so there are only a few surviving relics today. These include the masonic pillars with unicorns at the top of Middle Meadow Walk; the whale-jawbone arch at the southern end of Jawbone Walk (part of the display from the Shetland Isles); and the Prince Albert Victor Sundial, found in the Meadows itself.

Edinburgh's Dean of Guild, Sir James Gowans (1821–1890), a somewhat maverick architect and builder, was largely responsible for the exhibition's organisation. He had suffered serious financial losses in 1875 owing to overinvestment in the New Theatre in Edinburgh, which he was forced to sell in 1877 at one-third of its building cost. Although Gowans was knighted by Queen Victoria in 1887 in recognition of his contribution to the exhibition, which was itself reasonably profitable, he was bankrupted the following year when the Caledonian Railway obstructed a quarry extension at Redhall.

In the second half of the nineteenth century, many commercial mapmakers including Bartholomew did a brisk trade in various types of printed maps for railway companies. Sometimes these were quite ephemeral, created for particular events such as this one. The Caledonian Railway and its partner, the London & North Western, were keen for visitors to travel to the Edinburgh Exhibition via their trains to Princes Street Station (now the Caledonian Hotel); and although the North British Railway could have happily taken passengers too, via Waverley Station, Waverley is rather less helpfully marked 'For Caledonian Trains to the North'. Bartholomew's spartan map highlights all one needs to know to reach a single, specific, all-consuming goal: the 25-acre site and magnificent exhibition hall on the western side of the Meadows.

The family firm of Bartholomew traces its history over six generations back to George Bartholomew (1784–1824), who worked as an engraver under Daniel Lizars. George's son and grandson, John Bartholomew senior and junior, also trained as engravers and worked on a range of material including maps, usually for other Edinburgh publishers including Adam and Charles Black, William Blackwood & Sons, and W. & A.K. Johnston. Under John George Bartholomew (1860–1920; right) – who established the private limited company of John Bartholomew & Co. in 1888, and the public limited company of John Bartholomew & Son Ltd in 1919 – the focus shifted to making maps, and to publishing these under their own

An official portrait of John George Bartholomew (1860–1920), c.1900.

family name, rather than for other publishers. Under John George the firm gained an international reputation for high-quality cartography, combining high ideals and academic rigour with accomplished and innovative aesthetics, underpinned by state-of-the-art lithographic printing technologies. In the twentieth century under John 'Ian' Bartholomew (1890–1960), and his sons John C., Peter and Robert Bartholomew, the firm managed to retain its family ownership for much longer than Edinburgh's other family-owned publishers, who were steadily acquired by large multinational companies. Nevertheless, Bartholomew were eventually taken over by Readers Digest in 1980, and News International in 1985, finally deserting their Edinburgh home in 1995.

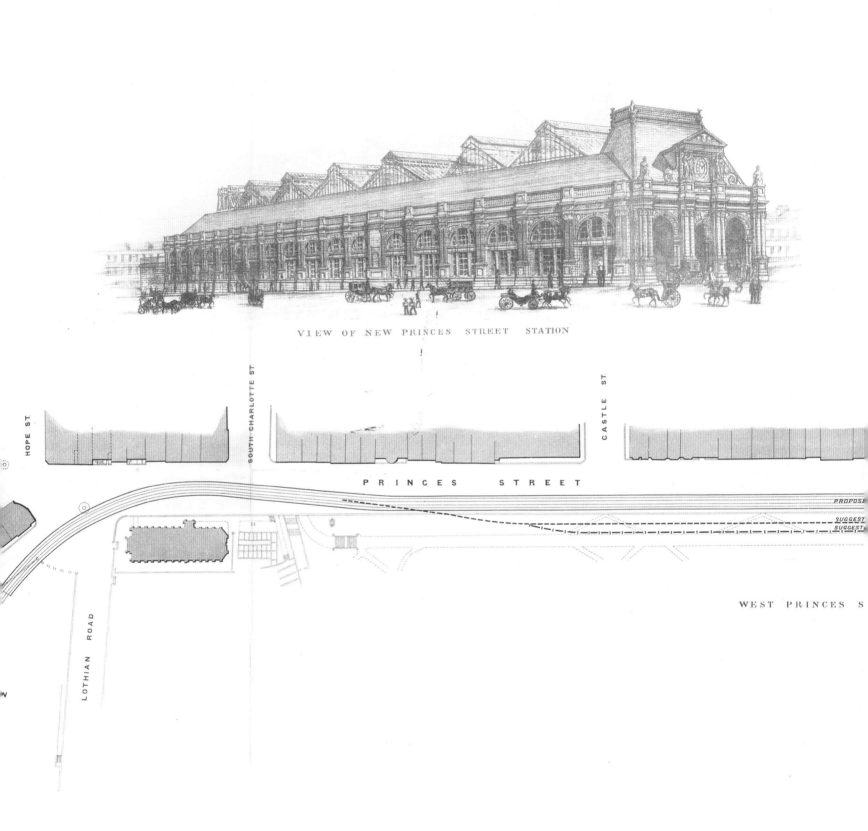

VIEW OF NEW PRINCES STREET STATION

1890

The Caledonian's tunnel along Princes Street

In 1890, the Caledonian Railway made an ambitious but ultimately failed attempt to extend their lines beyond their original Princes Street Station terminus to the North British Railway's Waverley Station. This was a particularly exciting time for railways, with the Forth Bridge having newly opened in March 1890; steadily growing passenger numbers, particularly of tourists; and a desire to extend local railways within Edinburgh, particularly to the northeast. The Caledonian were all too aware of the drawbacks of their original Lothian Road terminus, a temporary wooden structure, and in May 1870 opened a new station slightly closer to Princes Street. The company expended vast sums on the substantial demolition of buildings further north on Lothian Road and on the south side of Rutland Square, including St George's Free Church, the Royal Riding Academy, and the Scottish Naval and Military Academy, but the resulting station still lacked the centrality of Waverley and its connecting lines to the east and north.

The proposed new underground railway gave a choice of three minor deviations, indicated here with dashed lines, all of which involved tunnelling along most of the length of Princes Street, curving in and around the Royal Scottish Academy, and then curving in to what was then the Waverley Vegetable Market on the north side of the station. The cut-and-cover tunnel was planned to be just 4–5 feet below the road surface by the Caledonian Station itself, deepening to 6–10 feet from Charlotte Street eastwards. At the same time, the Caledonian brought in architects Charles Kinnear and John Peddie to design a handsome new neo-Baroque station with a triple-arched entrance and coupled Corinthian columns, which was built in 1892–93. Costing £120,000, the massive structure was 850 feet long and 190 feet wide on average,

---

Caledonian Railway, Proposed railway along Princes Street
and station at Waverley Market (1890)

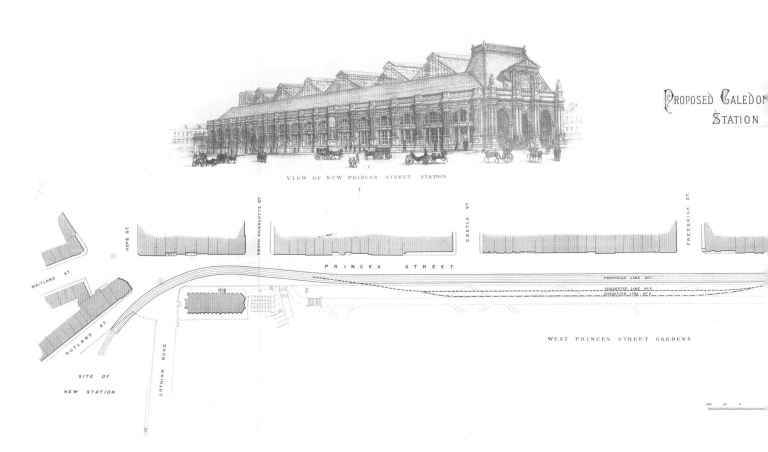

creating a covered area of 18,000 square yards, perspicaciously constructed so as to allow a hotel to be built over it in the future. Distinctive red sandstone, also used by the Caledonian for their stations in Glasgow, was brought in from the Corsehill and Corncockle quarries in Dumfriesshire via the Caledonian's own railway lines.

However, major objections to the planned underground railway were lodged by Edinburgh residents and the Town Council, particularly owing to the tunnelling process in Princes Street, and the plans were dropped in the early 1890s. However, the Caledonian Station itself was popular and successful, and was extended in 1899–1903 into a new hotel, moving even closer to Princes Street, with seven curved platforms and street-level access. This was a significant advantage over Waverley, and during the twentieth century, the Caledonian became the preferred arrival point in Edinburgh for members of the British monarchy on state visits. The entrance from Rutland Street to the west allowed access by carriages or cars, and also boasted its own booking office, left-luggage office and 'Refreshment Room'. Including its large goods and mineral yards laid out further south at Lothian Road and Morrison Street, the whole site occupied nearly 14 acres. The station was run down after nationalisation from 1948, and closed in 1965. Demolition in 1968, albeit with the retention of the hotel, allowed the substantial redevelopment (chiefly for financial institutions) of the adjacent area to the south.

The North British Railway Hotel, now known as the Balmoral, was constructed 1895–1903 in conjunction with the redevelopment of Waverley.

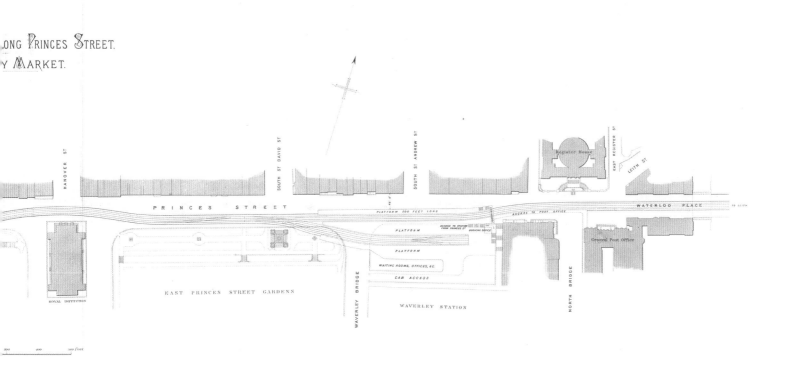

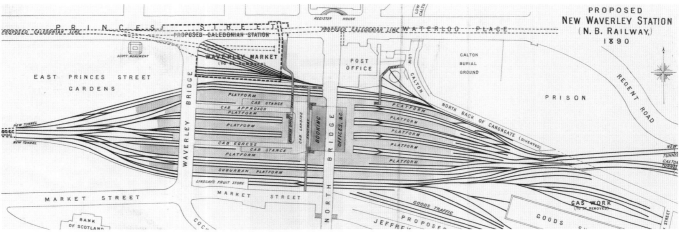

The North British Railway's *Proposed New Waverley Station* (1890), anticipating how the Caledonian Line could curve in from Princes Street through the demolished Vegetable Market. This never happened, but the expansion to the southeast, including the construction of East Market Street, was swiftly implemented.

1891a

'The finest and most elaborate map of the city and suburbs ever produced'

So boasted Bartholomew in their prospectus for this exceptional map, published on 12 sheets in the summer of 1891. Even allowing for publishers' puff, their claim had some justification; and, a century on, it is hard to avoid being struck by the incredible detail and superb aesthetics of this map. The Bartholomews were desk-based cartographers, and one of their greatest skills was in revising and repackaging information gained from other sources: in this case the Ordnance Survey large-scale town plan of 1876–77 [1852/1877]. Though the detail of the 66-sheet OS town plan would go unrivalled, even by the later work of the OS itself, its finely engraved uncoloured sheets were hardly eye-catching, and would have extended over some 24 × 18 feet if placed together. Bartholomew carefully reduced this to a much more manageable quarter of the size (the 12 sheets cover about 6 × 5 feet), brought it up to date, and enhanced the topography with subtle shades of green, blue and grey. The map also covers a similar extent on the ground to Kirkwood's [1817] and at a similar scale of 15 inches to one mile or 1:4,224, and so is particularly useful for illustrating the profound changes undergone by Edinburgh over the course of the nineteenth century.

The OS's original mapping, of course, allowed a wealth of detail: the interiors of public buildings, passageways, steps, pavements, flower-beds and paths in gardens, and even lamp-posts, shown with neat dots. Part of the skill involved in the creation of the present map was in consciously omitting other details to reduce clutter: gone are all the OS abbreviations for manholes, sewer-grates, water-taps and bench-marks, as well as most spot-heights above sea level, isolated trees, and the incredible tally of seat numbers inside the churches. John Bartholomew junior (1831–1893), who drafted this map,

---

John Bartholomew. *Bartholomew's Plan of the City of Edinburgh with Leith & Suburbs, Reduced from the Ordnance Survey and Revised to the Present Date* (1891)

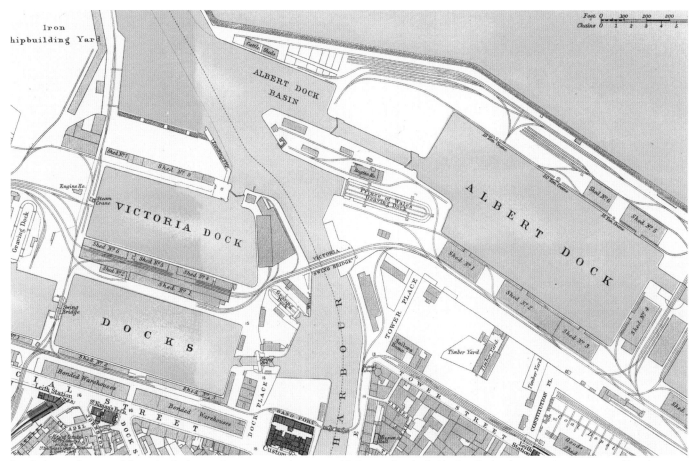

Bartholomew borrowed from the Ordnance Survey town plan to provide impressive dockside detail of railways and cranes, as well as the interior layouts of the Custom House and Leith Station. In the following decade the Imperial Dock would be constructed to the north, creating additional deep-water capacity.

trained for two years with the celebrated German geographer Augustus Petermann in the London offices of Justus Perthes of Gotha, and he was not only an expert engraver but also keen to experiment with the latest lithographic printing technologies. Copper-plate engraving lent precision to the black line-work and text, with features clearly distinguished through multiple fonts and excellent contrast, whilst subsequent multiple pulls through lithographic presses (three or four per sheet) added the colour. Work began on this map in 1880, and although John junior formally retired in 1888, he continued to work actively with his son, John George (1860–1920), most notably in preparing content for their monumental *Survey Atlas of Scotland* (1895). Only 500 copies of this map were ever printed, a small run by Bartholomew's standards: it occupied a specialist, utilitarian niche, compared with their more commercial, profitable maps that celebrated and elevated their home town.

This part of the map shows the docks at Leith, which had expanded almost beyond recognition in the previous 40 years, handling a shipping tonnage of nearly 1.5 million, about five times the 1853 volume. With the addition of the Victoria Dock (opened 1851), Albert Dock (1869) and Edinburgh Dock

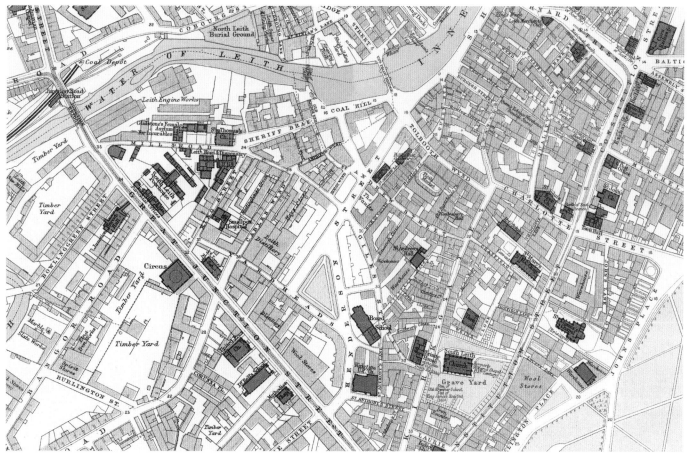

The Leith Improvement Scheme of the 1880s had opened up part of central Leith, driving Henderson Street through surrounding high-density housing and narrow, irregular streets. The darker shading for public buildings particularly highlights the growing number of new schools in the wake of the Education Act of 1872.

(1881), the overall pattern of docks and wharves came largely to resemble its present-day form; the Imperial Dock was authorised in 1892 and constructed north of the Albert Dock in the following decade. Within 'Old Leith', the most striking recent change was the implementation of the Leith Improvement Scheme of 1880. This drove the new Henderson Street (named after the then-Provost of Leith, Dr John Henderson) through several old closes, yards and houses, and displaced some 700 houses with 2,150 inhabitants. It is also worth noting the large number of new schools: the Education Act of 1872 had established local school boards and made education compulsory for 5–15 year olds, and by 1891 the Leith School Board administered 12 schools with an average of 1,000 pupils apiece.

The annual updating of the Post Office Directory maps would have been an obvious source for this and other new additions for Bartholomew. But other 'new' information came from the Ordnance Survey's own requirement to record antiquities – especially those with a military connection, such as the former Somerset's Battery and Pelham's Battery [1560] further south on Leith Links.

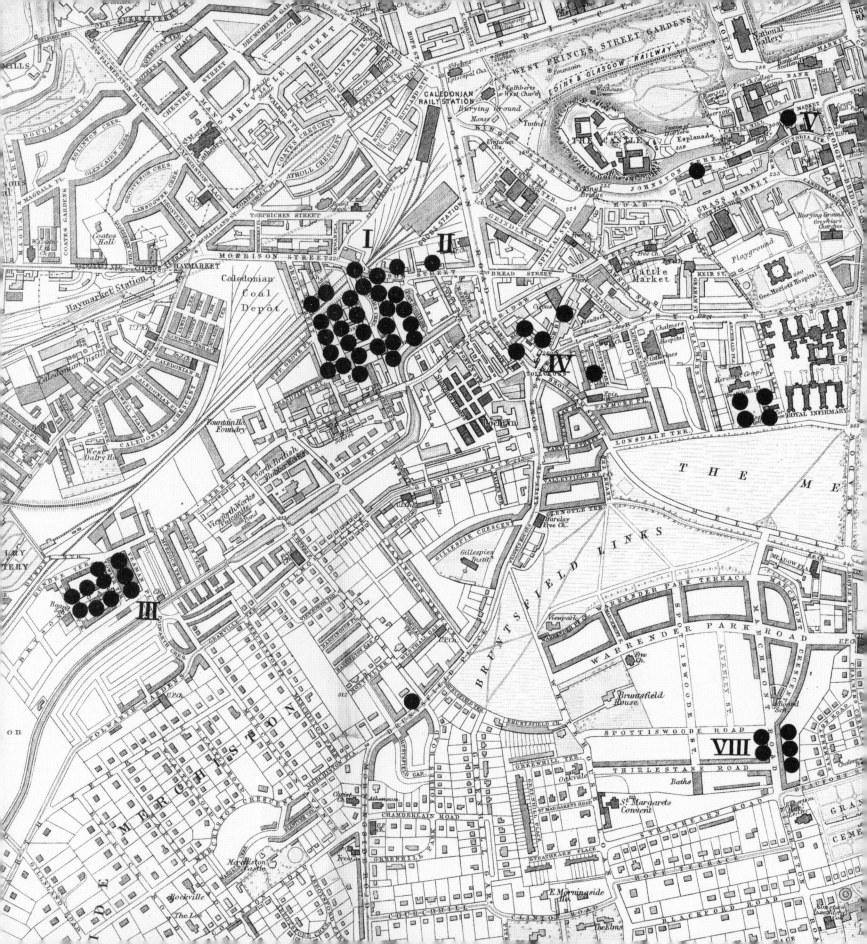

1891b

Mapping a typhoid outbreak to track down its source

This map was published in the *Edinburgh Medical Journal* in March 1891, accompanying a paper by Dr Harvey Littlejohn, which had also been delivered as a lecture to the Medico-Chirurgical Society of Edinburgh in January of the same year. Littlejohn had become well versed in Edinburgh's public health issues through his father, Sir Henry Littlejohn [1866]. At this time, the younger Littlejohn was a lecturer in Medical Jurisprudence and Hygiene in the Edinburgh School of Medicine, as well as Medical Officer of Health for Sheffield. He relinquished this latter post in 1897, in the first instance to assist his father, and later acted as Edinburgh's Chief Police Surgeon (1906–08) and Dean of Medicine and Professor of Forensic Medicine at Edinburgh University.

As Harvey Littlejohn himself wrote, 'no history of Edinburgh could be written without frequent reference to [typhoid] . . . at one time practically endemic in the Old Town', with significant mortality for both rich and poor. Symptoms began with lowered heart rate, nosebleeds, headaches and fever; this moved on to delirium, then intestinal symptoms – including, in the worst cases, perforation and haemorrhages. Death or recovery would generally occur within four weeks. During the nineteenth century, improvements in sanitation and living standards for most people had reduced its incidence very much to the poorest communities in the Old Town. Between 1880 and 1889, 355 cases were brought to the attention of the authorities, and 102 died; but in the following decade this fell to fewer than 80 cases and 19 deaths. Although the bacillus that was responsible for typhoid was discovered in 1880, it was only during the twentieth century that clean drinking water and effective vaccination properly brought the disease under control.

From an unexceptional background number of 16–30

H.H. Littlejohn / J.G. Bartholomew, *Map illustrating Dr Harvey Littlejohn's paper 'Distribution of typhoid cases'* (1891)

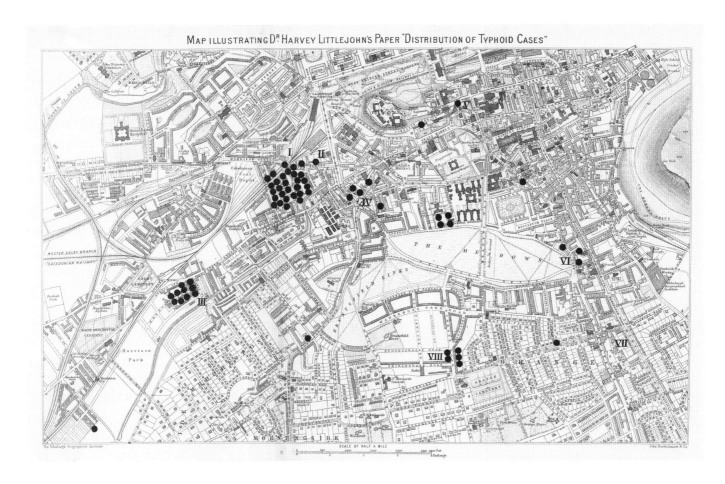

cases per month in January–September 1890, there were rapid increases – to 79 in October, 43 in November and 178 in December – and when these were plotted on a map they revealed significant clusters. Each Roman numeral shown related to a particular milk shop, and when information was aggregated, attention came to focus on three particular farms which supplied them. On inspection two were declared sound, but at the third the farmer's daughter and another relative were both infected with typhoid; further investigation revealed a well contaminated with sewage, a filthy byre, and milk cans washed in contaminated water. The clinching evidence came from Cluster VIII in Marchmont, where the same farmer delivered milk directly from his cart. Fortunately, the spread of the disease was halted by rapid action in stopping the supply of contaminated milk and making sanitary improvements at the farm, as well as in isolating and treating the victims where possible. There were only three deaths in total, and the number of reported cases of typhoid returned to normal by January 1891.

As well as clearly illustrating the power of mapping to identify the sources of disease (pioneered by John Snow's cholera mapping in London in 1854) and the importance of empirical evidence in trying to come to grips with a disease that was still not fully understood, this map underscores the period's keen interest in improving public health. Considered as a map of Edinburgh poverty in the generation immediately following the slum clearance of the mid Victorian period, it is also a telling record of the rapidity with which squalid condi-

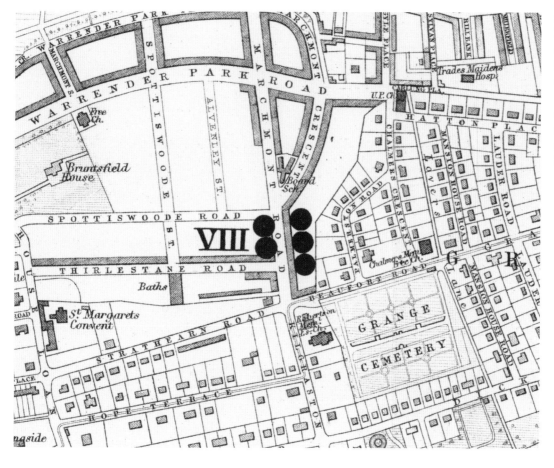

LEFT. Cluster VIII in Marchmont, where typhoid cases were traced to an infected farm's direct delivery by milk cart.

OPPOSITE. The clusters of typhoid cases were also traced to particular milk shops.

tions had migrated into newly built neighbourhoods: with only Cluster V (one of the smallest) located in what had been, until recently, the city's poorest and most shoddily constructed area.

Littlejohn's accompanying paper demonstrates remarkable intellectual rigour in its methodical examination of the facts and its discounting of various other possible causes of the outbreak, and reaches a qualified but firm conclusion based on detailed personal observation. The author had clearly spoken to many patients, visited their houses, and spent time investigating the infected farm, even drawing a plan of the buildings and well. However, he was careful not to name it, given that it was the farmer's only livelihood and that there was no medical reason he should not continue in business, provided that appropriate improvements were made. Littlejohn also ensured that shopkeepers who had been forced to dispose of the contaminated milk were offered compensation. In another paper of 1899 co-written with Claude Ker, superintendent of the City Fever Hospital in High School Yards, he was quite clear that poverty was the major cause of typhoid in Edinburgh, but not only poverty. 'Even those residing in the same tenement and living under conditions most favourable to receive infection, namely dirt, squalor and intemperance, do not become infected unless they have been in close contact with the patient.' Public health would be improved through proper evidence and an open mind, not just a moral crusade.

1892a

Tuberculosis as a window on late Victorian Edinburgh society

In late-nineteenth-century Edinburgh, consumption (tuberculosis) was an even graver threat to health than typhoid was, being responsible for some 10 per cent of all deaths in 1890. Poorly understood, the disease was popularly blamed on everything from prostitution to vampires, but even the well-educated tended to see it as a fellow-traveller of vice and sin. The bacillus that actually caused tuberculosis had been identified and described in 1882 by Robert Koch (1843–1910), who also isolated anthrax and cholera and went on to win the Nobel Prize for Medicine in 1905. However, the first widespread use of an effective vaccine occurred only after the Second World War. In the interim, dispensaries and sanatoria made important contributions to therapy, but more widespread improvements in hygiene and public health were also important, as physicians struggled to understand precisely which environmental factors gave rise to the condition.

In 1887, Dr Robert Philip (1857–1939), who had studied Koch's work in Vienna, founded the Victoria Dispensary for Consumption and Diseases of the Chest. Initially located in a set of flats in Bank Street, Edinburgh, it moved to 26 Lauriston Place in 1891. This dispensary was the first preventative institution of its kind in the world, and particularly valuable for the poorer inhabitants of the city. As a core component of the Edinburgh Anti-Tuberculosis Scheme, it would also have a seminal influence on the development of similar models in other towns, but not before the critical intervention illustrated here.

This map by Philip plots the locations of the first 1,000 cases of tuberculosis recorded by the new Victoria Dispensary during 1887–90. It is interesting not only for how widespread the cases were, but also for their uneven concentrations. The worst-affected areas were the Old Town and South Side, and

R.W. Philip / J.G. Bartholomew, *Map of Edinburgh Showing Cases of Pulmonary Tuberculosis Received at the Victoria Dispensary for Consumption and Diseases of the Chest during Three Years* (1892)

westwards to Gorgie, in strong contrast to other outlying suburbs and the New Town which had just a small scatter of cases. Bartholomew used their standard Post Office Directory map [1934] as a backdrop, and the numbers on the map relate to Municipal Wards, with the lines of black dots indicating their boundaries. The map formed part of a detailed paper presented by Philip to the Medico-Chirurgical Society of Edinburgh in March 1892, synthesising vital statistics from the cases he had examined, and it was subsequently published in the *Edinburgh Medical Journal*.

Philip's table listing the occupations of these 1,000 patients contains some surprises; less than 5 per cent of the sufferers were children, while many others were *not* confined to poorly-ventilated environments or lower-income occupations, but represented a fairly wide cross-section of Edinburgh's domestic economy. Moreover, this economy itself – probably due to the city's historic status as a royal capital and centre of the legal profession – was not nearly as dominated by manual labour as other British cities of its size. At least 15 per cent of the victims would be definitely describable as lower middle class or above, even if we do not include the 71 in publishing-related occupations or the 69 who were employed or self-employed in the clothing trades, which would swell this figure to 29 per cent. At least some of the 141 married housewives – the single largest occupational category affected – were also presumably middle class in origin. And, of the minority of the victims (45 per cent) who were unarguably in working-class occupations, more than one-third (17 per cent) would have worked mainly outdoors. Philip was able to record the results of treatment for 469 of the patients: around 58 per cent were demonstrably 'improved', but 24 per cent were described as 'indifferent', and 16 per cent had deteriorated or died.

As Philip noted, Edinburgh's medical provision as of 1892 was completely inadequate to the task of treating or isolating tuberculosis patients, with many sufferers being confined to their homes. 'If there be the slightest degree of truth in the contagious view of tuberculosis, such chronic foci of infection ought not to be permitted to smoulder under conditions which are calculated to encourage the fatal propagation.' Edinburgh Town Council had used the Canongate Poorhouse as well as the City Poorhouse for epidemics, but these were unsuitable, and Philip and Henry Littlejohn campaigned for a proper isolation hospital, which was only successful in the construction of the Fever Hospital (later the City Hospital) in Craiglockhart in 1897.

The Victoria Dispensary and the Fever Hospital would both become key components of the Edinburgh Anti-Tuberculosis Scheme, which brought together diagnosis, referral, isolation and sanatoria environments alongside developing municipal anti-tuberculosis institutions. Patients presenting themselves were placed in one of three categories: advanced cases, early onset, and cured patients who required isolation or further recuperation in 'working colonies'. Advanced cases were initially sent to the City Fever Hospital. Early-onset cases were sent to a sanatorium, usually the Victoria Hospital for Consumption (VHC) in Craigleith (later the Royal Victoria Hospital), which was opened in 1894. Funded by public subscription, it grew from 15 beds to more than 100 over the next 13 years. In the VHC and similar institutions, it was hoped that fresh air and good hygiene would arrest the further development of the disease. Those requiring isolation and rest were often sent to Polton Farm Colony, opened near Lasswade in 1910, where patients could engage in gentle agricultural work in an environment very different from their homes. The Dispensary also took active measures to research and make records of the disease, organise home visits by physicians, train nurses and coordinate aftercare. All became part of the City-run Royal Victoria Tuberculosis Trust in 1914.

The death rate from tuberculosis in Edinburgh fell dramatically in the twentieth century, albeit with further peaks during the two world wars. It was 1.9 per 1,000 in 1887, but had fallen to 1.07 per 1,000 by 1910 and to 0.8 per 1,000 ten years later. Philip was knighted in 1913 and appointed physician to King George V.

MAP OF EDINBURGH
Showing Distribution of Cases of Pulmonary Tuberculosis received at the Victoria Dispensary for Consumption and Diseases of the Chest during Three Years.

The red dots indicate approximately the residence of the patients. The streets or divisions of streets have been marked with care. For present purposes it has been deemed unnecessary to localise the cases at particular numbers.

Dolphin Head
Black
Middle Craigs
Rocks
Eastern Craigs
Outer Light House
Light Ho.
Newhaven
Granton Point
Birny Rocks
Granton Ho.
Broomfield
Muirhouse
Granton
Granton Mains
Trinity Grove
LEITH
Leith Links
Station
Silverknowes
Muirhouse Mains
Pennywell
Wr. Pilton
Ed. Pilton
Drylaw Mains
N. Inverleith Mains
Rl. Botanic Garden
Davidson's Mains
Corbyhall
House of Hill
Drylaw Ho.
Easter Drylaw
Inverleith Ho.
Fettes College
Poorhouse
Craigcrook
Blackhall
Craigleith
Stockbridge
EDINBURGH
Stewart's Hosp.
Dean Park
Ravelston Ho.
John Watson's Institution
Murrayfield Ho.
Donaldson's Hosp.
Kirkton Lodge
Cathedral
Drum Ho.
Lochend Loch
Hawkhill
Restalrig
Fillyside
Piershill Barracks
Holyrood Palace
St. Anthony's Chapel
Willow Brae
Northfield
Meadowfield
Dunsapie Loch
Woodlands
Duddingston
The Meadows
Bruntsfield Links
Newington
Salisbury Green
Echo Bank
Prestonfield Ho.
Peffermill
Newington Cem.
Cameron Ho.
Niddry Mains
Craighall
Roseburn Ho.
Meadow Ho.
Carrick Knowe
Saughton Hall
Saughton
Saughton Mains
Gorgie
Gorgie Mills
Stenhouse Mills
Morningside
Blackford Ho.
Blackford Hill
Liberton W. Mains
Nether Liberton
Liberton Dams
Plewlands
Craig House
Braid Cottage
Craiglockhart
Craiglockhart Ho.
Slateford
Kingsknowe
Redhall
Hailes Cottage
Hailes Ho.
Wester Hailes
Colinton Ho.
Colinton Mains
Colinton
Greenbank
Oxgangs
Comiston
Redford
Spylaw Bank
Viewfield
Juniper Green
Woodhall
Bonally
Dreghorn Castle
Woodhead
Mounthooly
Liberton
Liberton East Mains
Alnwickhill
Backsidelee
Howdens Hall
Gracemount
S. Catherines
Morton Hall
Burnhead
Broken Bridge
Fernieside
Stenhouse
Little France
Bridge-end

1892b

Mapping goes underground

This map celebrates the great age of Edinburgh geological research under Archibald and James Geikie and the importance of mapping under the ground for understanding life above it. Edinburgh's underlying rocks and their histories define the spectacular topography of the city, and have dictated the patterns of settlement and land-use in more recent times. The 'Seven Summits' are all the result of volcanic activity in the Carboniferous and Devonian eras, 300–400 million years ago: Arthur's Seat is the remnant of a Lower Carboniferous volcano; Salisbury Crags is a large volcanic sill of hard intruded magma between beds of softer sedimentary strata; Calton Hill is a displaced fragment of the same volcano; and Edinburgh Castle is on a plug formed from one of the vents. Turnhouse Hill and Corstorphine Hill to the northwest are sills of volcanic dolerite, whilst the Pentlands, rising to the south, contain Devonian lavas and tuffs.

On the map, these volcanic intrusions are usually shown in red and orange, contrasting with the large swathes of blue and pink sandstone from the Carboniferous era, which were quarried extensively for the fine architecture of the New Town. Several thin white fault lines can be seen, the most important being the Pentland fault: part of the northerly section of the Southern Uplands fault, which traces a line from southwest to northeast from just below Liberton to the coast at Portobello. To the southeast are important light blue bands of limestone around Gilmerton and Cousland, while the dark grey around Dalkeith and Inveresk indicates coal, mined extensively in former times.

During the last Ice Age, scouring of these rocks by large glaciers sweeping eastwards created their distinctive current forms, with the crag-and-tail outline of the major hard volcanic plugs, interspersed with gouged areas such as the Nor' Loch and Grassmarket on either side of Edinburgh Castle. Glacial meltwaters also left substantial deposits of

---

Geological Survey of Scotland, *One-Inch to the Mile, Scotland. Sheet 32 – Edinburgh* (1892)

boulder clay (shown here in paler tones of beige), raised beaches, and extensive lochs. Several of the lochs have been drained and used in the last century for golf and race courses as well as sports grounds: Murrayfield and Hearts are both sited on the bed of Corstorphine Loch, while Hibs and Meadowbank take advantage of raised beach flats in the north of the city. The former Canonmills Loch was used for mills for centuries and appears prominently on Bartholomew's Chronological map [1919], whilst the loch flat at Turnhouse was used from 1915 as an aerodrome for the Royal Flying Corps, latterly named RAF Turnhouse and now Edinburgh's international airport.

The Geological Survey of Great Britain and Ireland, initially a branch of the Ordnance Survey, was created in 1835, and Sir Henry De la Beche (1796–1855), a close friend of pioneering Dorset palaeontologist Mary Anning, became its first director. He was followed by a succession of Scots: Roderick Impey Murchison, appointed in 1855; Andrew Ramsay in 1871; and Archibald Geikie in 1882. Although the main British headquarters were in London, with only a small temporary store for maps and specimens at the Industrial Museum of Scotland in Argyle Square, the Scottish branch steadily grew over time. From these initial premises they moved to offices in the India Buildings in Victoria Street (1869–79), then to the Sheriff Court Buildings on George IV Bridge (1879–1906), and eventually to George Square (1926–28). The organisation was able to follow cartographically in the wake of Ordnance Survey's progress northwards in Scotland from the 1850s. OS maps at 6 inches to the mile, with reductions to the 1 inch to the mile scale (as shown here), formed the main base mapping for plotting geological information. The Geological Survey maps were accompanied by detailed textual 'Memoirs', which were usually major written works describing the detail of the principal rocks depicted, including superficial glacial deposits, and their economic potential.

Edinburgh also boasts an important history of geologists and geological research. James Hutton, the acknowledged founder of modern geology, lived at St John's Hill [1784]. He famously discovered the geological unconformity at Siccar Point on the Berwickshire coast, while Hutton's Section – located on the flanks of Arthur's Seat – was where he gathered sufficient evidence to conclude that the dolerite sill of Salisbury Crags had been forced into the sedimentary layers in a molten state. A century later, the Edinburgh area was the initial research focus and base for the Geological Survey of Scotland, and the first OS one-inch geological map brought out in Scotland was Sheet 32 of Edinburgh environs, published in 1859 with fieldwork by Andrew Ramsay, Archibald Geikie and Henry Howell. Over the next half-century, Edinburgh geologists led the way in researching and mapping Scotland's geology, contributing to major international advances in geological theory. Perhaps foremost amongst these were the Geikie brothers, Archibald and James, who between them occupied the prestigious Murchison Chair of Geology at Edinburgh for 43 years. In 1882, when Archibald went to London to become Director-General of the Geological Survey, his younger brother James succeeded him as Professor.

In the 1860s, the Geological Survey saw a large increase in staff, following the recommendations of a Royal Commission on national coal reserves, and significant further work was undertaken on the Edinburgh environs by Ben Peach, James Wilson and Henry Caddell, whose names all appear at the lower left of the sheet. Although this map in essence shows a similar colour distribution and typology of rocks to its 1859 first edition, advances in geological knowledge allowed greater detail and more subtle classification differences to be brought out, building upon the initial survey work. Whilst the rocks might have always stayed the same, the understanding and representation of them over time on maps has steadily evolved, and present-day Edinburgh geological maps have even more categories and colours.

The Geological Survey of Great Britain remained part of the OS establishment until 1965, when it was detached and merged with London's Geological Museum to become the Institute of Geological Sciences. The resulting entity has been known as the British Geological Survey since 1984.

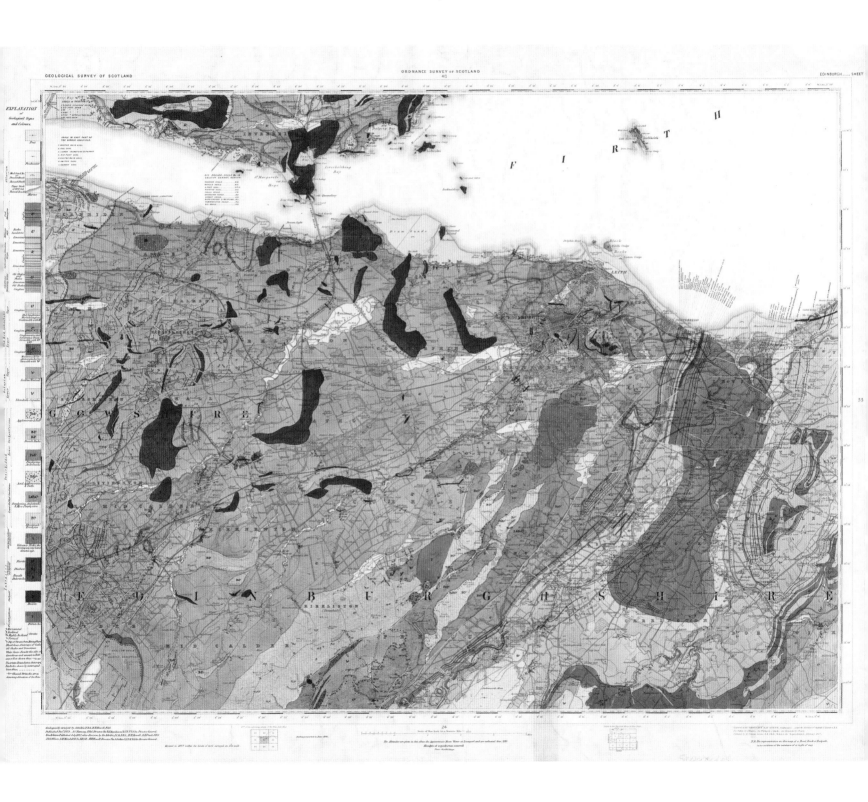

1895

From city to conurbation: powering and extending the tramway network

Perhaps more than any other single factor, a large and efficient tram network abetted Edinburgh's growth into a conurbation, with suburban infill joining the old city solidly to other old communities such as Leith, Musselburgh and Liberton to form a single entity. It is probably no coincidence that the term 'conurbation' was coined in 1915 by visionary urban planner Sir Patrick Geddes (1854–1932) [1902], who cited electric power and motorised transport as root causes of the phenomenon.

Following the Edinburgh Tramways Act of 1871, tram lines were constructed along certain streets by the Edinburgh & District Tramways Company. Like the 'Innocent Railway' [1851] and similar lines of the earlier nineteenth century, the double-decker trams were horse-drawn, with the rails serving to ease the work of the horses. A team of two drew each vehicle, though on hillier routes a third horse was added in front. Contemporaneous attempts to introduce steam omnibuses, each capable of carrying 50 passengers, proved unworkable; but the north–south routes through the New Town (shown here in blue) were too steep for horses, so cable tramways were installed there from 1884 by a second company, the Northern Tramways Corporation. These 'cable cars', conceptually almost identical to those installed in San Francisco from 1873, operated by gripping downward onto a constantly moving cable, housed in a shallow trough between the tram rails and powered from a building on Henderson Row (shown here as a blue rectangle). This is partially preserved today in the Scottish Life Assurance Office, along with winding gear outside.

This map captures the city's tramways at a time of transition, from private into public ownership, and in terms of how they were powered. During the 1890s, the Edinburgh Town

---

J. Bartholomew & Co., *Edinburgh street tramways.*
Routes proposed to be cabled (1895)

Corporation acquired parts of the system from the two companies in stages (particularly in 1894–95), and began a programme of expansion and further mechanisation. Whereas in 1893 there had been 1,100 horses at work, the extension of the cable network led to a steady decline, and the last horse-drawn tram ran in 1907. Two new cable-winding stations were also set up, as indicated here: one at Shrubhill just off Leith Walk, and a second at Tollcross, immediately to the west of the Central Methodist Hall. The cable tram system was generally effective and grew to become one of the largest in the world, extending over some 36 linear miles; but its speed of travel was underwhelming (as the nickname 'cable crawl' would suggest), and a single broken cable could reduce the entire system to a standstill.

The Corporation therefore looked towards electrification of the system from the 1920s, an initiative spurred on by its takeover of Leith's tramways, which (like Musselburgh's) had converted to electric traction as early as 1905. The different systems had meant that passengers travelling up or down Leith Walk had to change trams at Pilrig on the boundary between the two towns, resulting in a confusing interchange of passengers known locally as 'the Pilrig muddle'. Once it got going, the conversion to electric traction was rapid, and the last outing of a cable-powered tram took place in June 1923.

The colours and symbols for the tramways are slightly confusing: red dashed lines indicate proposed new cable-powered routes, whilst black dashed ones indicate existing horse-drawn routes that were not proposed for immediate cabling. Proposed routes shown here, but which were not actually developed, included G–H from Claremont Street to Bonnington Road, K–L along Easter Road, and W–X through Middle Meadow Walk to Warrender Park Road. However, the proposed extensions south from Morningside and Cameron Toll were built, eventually extending as far as Fairmilehead and Liberton respectively. There were also extensions, not shown here, from Gorgie to Stenhouse (1930) and North Gyle to Maybury (1937). The effects on housing development were profound indeed: not long after electrification, it would have been theoretically possible for Tarzan to move between Canongate and Levenhall, a distance of 9 miles, by swinging from rooftop to rooftop.

By 1939 the electric tramways extended over 47.5 miles – which was more than bus services, despite the latter's steady expansion throughout the interwar period. However, further extensions were put on hold during the Second World War. Thereafter, the system was not maintained and steadily lost support in comparison to other means of transport – including, especially, newly affordable private cars [1949]. Many routes were abandoned in the early 1950s, and the last tram ran in 1956. However, a new era is now upon us, following the controversial and troubled 2007–14 construction of an 8-mile tramway from Edinburgh Airport to York Place.

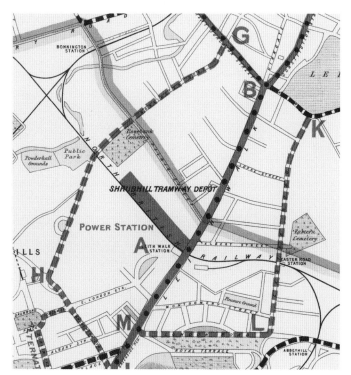

The Shrubhill Tramway Depot and Power Station was opened in 1898 to house haulage engines for the cable tramways, but many of the proposed new cable routes shown with dashed red lines were not implemented.

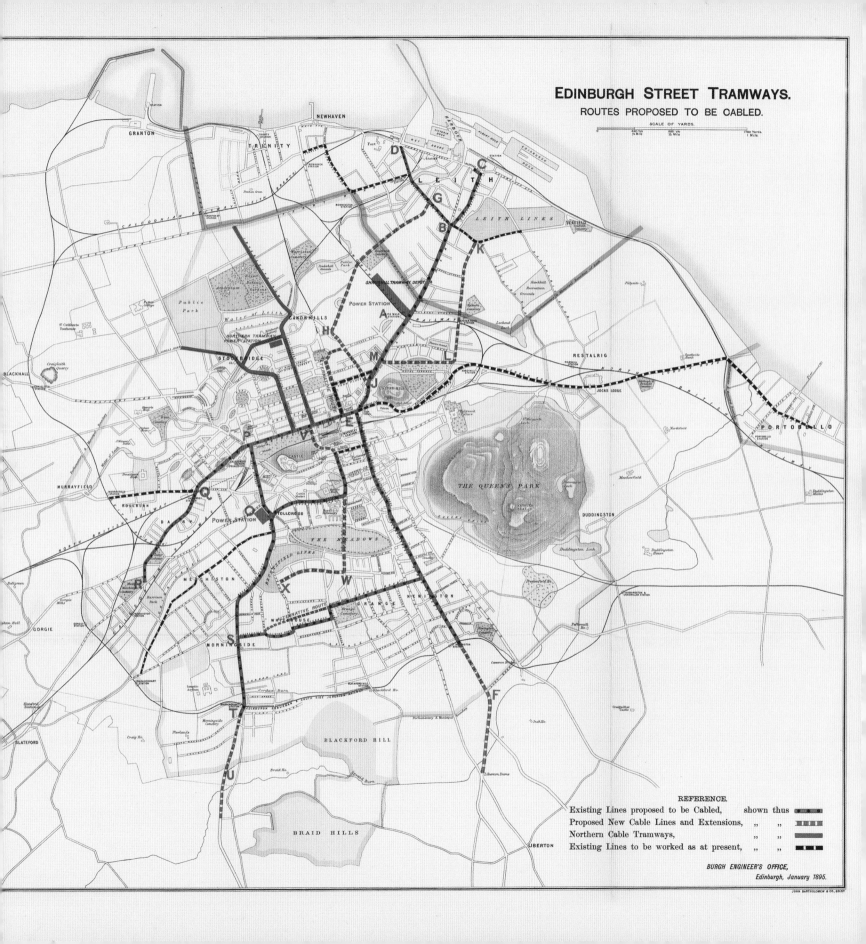

1898a

Advancing into the light

Edinburgh has a long history of lighting its streets: initially using whale oil, then gas, following the establishment of the Edinburgh Gas Light Company in the north back of the Canongate in 1817. This map shows the progress of electric street lighting from its inception in 1895 until 1898.

Lanterns were hung outside the town's main ports from an early date, albeit intermittently. Following a royal charter of 1688, however, more concerted attempts were made to light the High Street properly, and the first public lamplighter was employed beginning in 1701. From 1785, Police Commissioners were appointed and given responsibilities for lighting and cleaning; by the following year, more than 300 lamps were in use, mainly wall brackets, and this number grew exponentially as the construction of the First New Town proceeded [1819b]. By 1820, there were 4,781 lamps in operation, 4 per cent of them gas-powered. In spite of the high numbers, the quality of the lighting was widely criticised: one wag proclaimed that it provided 'light just enough to make part of the darkness visible'.

British towns rapidly electrified their street lighting in the later nineteenth century. Edinburgh had been one of the first in the world to adopt gas, years ahead of Paris, Berlin, Washington and St Petersburg. Yet when work began on electrification in the 1890s the Town Council were, at best, able to boast that they had been able to learn from other towns' experiences. Recent major difficulties they had suffered in seeking to acquire water and gas supplies from private companies encouraged them to promote a municipal scheme from 1893, with the Town Council in full control. Work began properly from April 1895, with the electricity coming from a specially built power station in Dewar Place off Morrison Street, handy for fuel deliveries via the adjacent Caledonian

---

Edinburgh Town Council / J.G. Bartholomew,
Street Lighting: Plan of Mains (1898)

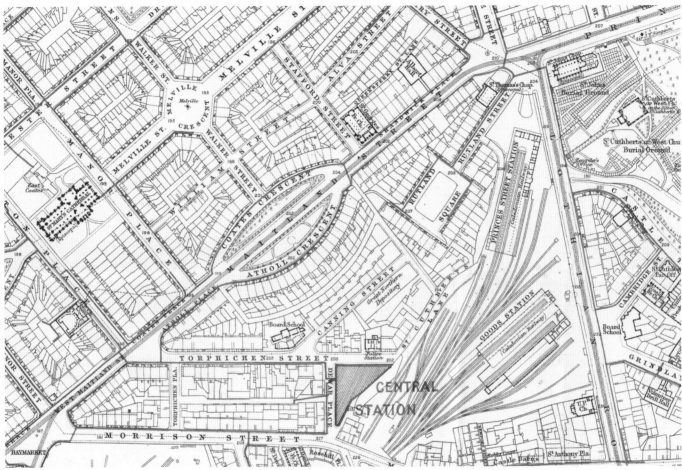

The new electric street lights were powered from a specially-built station in Dewar Place off Morrison Street. Some of the New Town and West End streets were lit on both sides.

Railway. Growth was rapid thereafter: in each of the following three years, capacity equivalent to around 50,000 8-candle-power lamps was added; by 1898, some 1,650 private houses were connected, and the system extended over 20 miles. Even reading between the lines of the council's hyperbole, the initiative had exceeded expectations economically, contributing over £4,000 to a reduction in rates, in spite of (or because of) being one of the cheapest systems in the country.

That said, not all Edinburgh residents approved. The digging-up of streets and pavements was a temporary nuisance, but much more serious and alarming were the frequent explosions caused by escaping gas fusing with the electric wiring under the streets. On occasion, residents were shocked to see paving stones blown out of the ground, followed by flames 3–4 feet high, hissing noises and the fading of nearby lamps. Some were also unhappy about the large quantities of black smoke belching forth from the power station, and expressed nostalgia for 'the clear skies and fine sunsets we used to enjoy'. This problem multiplied in 1898 when the new MacDonald Road power station was constructed to meet escalating demand.

There were also criticisms of the poor location of lights,

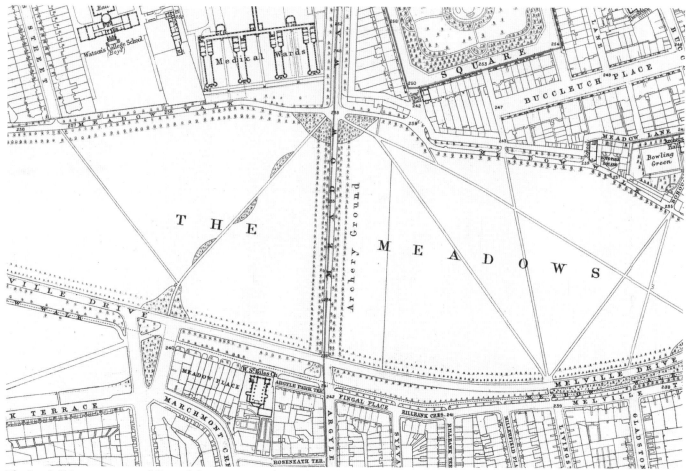

The South Side was not so well lit, and there were complaints about the lighting running directly down the centre of thoroughfares, including Middle Meadow Walk.

often in thoroughfares or otherwise causing obstructions: for example, running straight down the centre of Middle Meadow Walk. As a writer to the *Scotsman* noted in September 1897, these caused a great inconvenience to the public as well as danger to the blind, being 'proof of a great want of light among civic officials of that department'. Perhaps most fundamentally, as illustrated here, the lighting only covered very limited areas in the centre of Edinburgh and along certain major thoroughfares out of the city. Some streets in the New Town and West End were lit on both sides, whilst streets not far away were entirely missed. By 1903, electric lighting covered at least 38 miles of streets, but this still left some 150 miles lit with inferior gas light, and some not lit at all, with little sustained effort made to improve the situation further.

Bartholomew printed 200 copies of this map for the Town Council on 11 February 1898, very much as an in-house tool to illustrate the progress of electrification as of that date. The base map was Bartholomew's fine large-scale *Plan of the City of Edinburgh with Leith & Suburbs* [1891a], forming an excellent scale with good detail for the overlay. The council were also supplied with blank copies, presumably for sketching in and planning further extensions.

1898b

The Usher Hall on the Meadows

When the well-heeled founder and chairman of the North British Distillery Company, Andrew Usher, kindly donated £100,000 in 1896 to fund a hall 'for concerts, recitals or other entertainments or performances of a musical nature' one of the only stipulations in his gift was that it should be 'begun forthwith, in order that he might have the pleasure of seeing it reared in his lifetime'. He could hardly have expected it would take 14 years just to find somewhere to put it. Unwisely, as it turns out, the Town Council invited the good citizens of Edinburgh to contribute their ideas for the most appropriate site, thus unleashing an endless torrent of suggestions in the local press as everyone had a try at pinning the Usher Hall tail on the donkey. Serious dilemmas, quite familiar to us today, were also raised: should good buildings be demolished, or green space encroached upon? Should the hall be sited in the usual concert and theatre zone (then the New Town), or should it be used to initiate a new cultural hub elsewhere? Should it be used for civic improvement or as part of a wider planning scheme?

The top site from this last perspective was definitely the Canal Basin with its surrounding maze of breweries, slaughterhouses and tenements. But its owners, the North British Railway, were not keen to part with the site; and as one objector noted, expressing a wider sentiment: 'one would hardly care to plant Versailles in Fountainbridge'. Another innovative suggestion was to situate the new structure in front of Atholl Crescent, with traffic in Maitland Street / Shandwick Place circling around it like St Paul's. The southwest corner of Chambers Street was also a serious contender. However, the former was rejected by both Houses of Parliament, whilst

---

Edinburgh Town Council / J.G. Bartholomew,
*Plan of the Meadows and Part of Bruntsfield Links shewing the
Site Proposed for the New City Hall (Usher Hall)* (1898)

the latter would have required demolition of good buildings. The Chambers Street site would not be built on until 1999, as part of the National Museum extension.

In 1898, the Meadows emerged as frontrunner, a choice perhaps encouraged by the opening of the People's Palace on Glasgow Green in January of that year. The scheme had the immense advantage of requiring neither expensive demolition nor arbitration with the owners, being completely in the council's hands. And the Usher Hall would, after all, take up only three-quarters of an acre in one corner, leaving 74 acres for recreation and amenity. (The cricketers could surely be offered compensation?) Fatally, however, an Act of Parliament from 1827 had definitively barred permanent construction on the Meadows [1886], and though some members of the Town Council thought that this could be repealed, others were doubtful, and neither faction doubted that it would be a tough battle.

The council commissioned 100 copies of this map, which were printed on 5 July 1898 in preparation for a key vote on the Meadows site two days later. Eagle-eyed readers may recognise the underlying map as an outline version of Bartholomew's fine 1891 plan [1891a], with the obvious addition of the proposed hall itself. It would be difficult to argue that the map favoured one side or the other; but in any case, after months of feuding, the council rejected the Meadows site proposal by 27 votes to 10. Somewhere else now had to be found. Less than four months later, Andrew Usher passed away, having lived just long enough to see his dream descend into a vaguely comical nightmare.

The council increasingly set its sights on Castle Terrace, and in 1903 took the opportunity to purchase the Synod Hall (1875; demolished 1966), which was available owing to the 1900 amalgamation of the United Presbyterian Church and the Free Church of Scotland. This explains the appearance of this site on our slaughterhouses map [1903a], the hope having been that the Usher Hall could – in addition to its main functions – absorb the role that the Corn Exchange had played as a venue for political meetings. However, there were numerous objections to the proposed demolition of the Synod

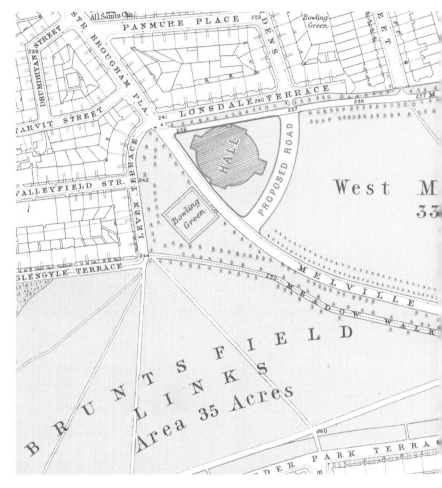

Hall, and it was only in March 1910 that the final site fronting onto Lothian Road became viable, following complex negotiations with the School Board, among others. Unbelievably, even the appearance of the building had not yet been decided, and this was now made the subject of an architectural competition. The Leicester-based firm of Stockdale, Harrison & Sons beat the competition, and construction started in 1911, with the first concert being performed three years later.

The decision to allow political meetings in the Usher Hall caused a number of problems which came to a head in 1934, when Oswald Mosley came to speak. Preaching his policy of

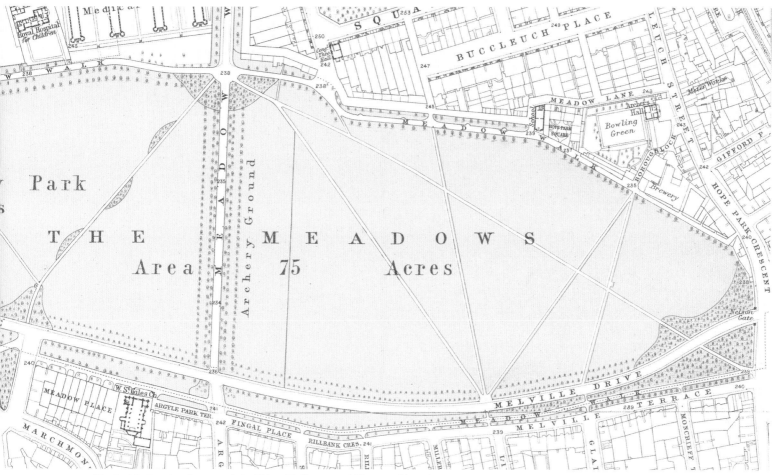

The legislation that frustrated the Usher Hall proposal illustrated here has helped to keep the green space of the Meadows much the same today.

'British first, the Dominions next, and the foreigners nowhere', the British Union of Fascists' (BUF) leader was picketed both by Scottish Nationalists and Communists, who fought with his 300-odd uniformed bodyguards. A crowd of several thousand gathered and there were some dozen casualties, of whom more than half were 'Blackshirts' and the rest mostly innocent bystanders including a bus driver. Nevertheless, Mosley's speech was praised in the *Scotsman* newspaper as 'fluent' and 'logical'. BUF propagandists, stung by the strength of Scottish Nationalist opposition to their message, later wrote that only fascism could 'restore the loyalty of Scotland to the Union'.

Incredibly, after he was thrice refused permission by the Town Council in the winter of 1935–36, Mosley was allowed to return to the Usher Hall by a 'specially convened' meeting of the full council, and did so on 15 May. Of course, nearly the same thing happened again, albeit with the violence concentrated inside the hall during the meeting rather than outside afterward. This time there were a dozen arrests, and the newly anti-Semitic content of Mosley's speech led his own propaganda officer, Bryham Oliver, to resign. Further violence and arrests occurred at a pro-Franco meeting in Usher Hall in June 1938.

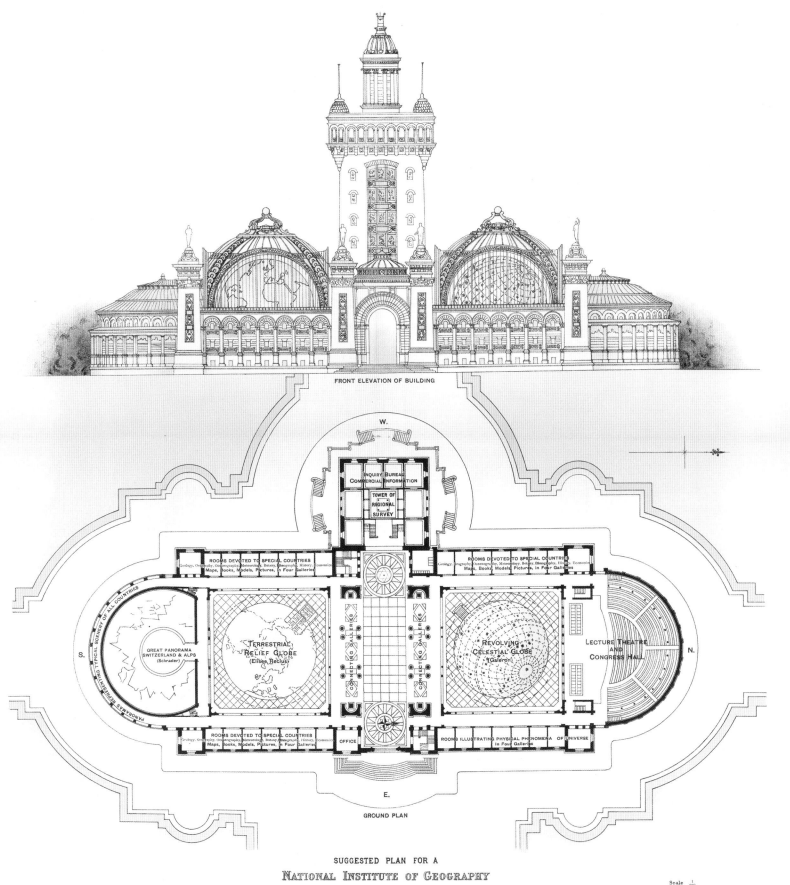

SUGGESTED PLAN FOR A
NATIONAL INSTITUTE OF GEOGRAPHY
According to Designs of Professor Geddes.

1902

Civic survey, social evolution and the Encyclopaedia Graphica

This stunning plan and elevation of the proposed National Institute of Geography in Edinburgh is important as an expression of several key ideas of Patrick Geddes, who stressed the importance of regional survey and geography for social evolution, and the relationships between the local community and its physical environment. The Institute was also an *Encyclopaedia Graphica* of the visualised universe – a method for systematically arranging and exhibiting the world's knowledge based on a geographical model. Geddes had a critical impact on the development of India (where he taught sociology for five years), Palestine (where he was commissioned to re-design the city of Tel Aviv), and his native Scotland; and he is considered a founding father of town planning and urban regeneration around the world. The widely used environmentalist slogan 'Think Globally, Act Locally' was paraphrased from Geddes's book *Cities in Evolution* (1915).

Geddes firmly believed that to understand and improve an area one had to be part of it; and in 1886 he and his wife acquired a row of slum tenements in James Court, just off the north side of the Lawnmarket. He set about renovating these, and founded the Environment Society (later the Edinburgh Social Union) to encourage other local residents to survey, plan and improve the local environment through a project of 'conservative surgery': those buildings past repair should be pulled down, but those that were repairable should be saved. The gaps left by the demolished properties were to be made into courtyards and small gardens, and many of these – either created in Geddes's time or inspired by his ideals – survive in the Old Town today. Geddes hoped that this would encourage an influx of more affluent residents, allowing the development of a mixed social structure and a healthier community.

In the early 1890s Geddes moved to Ramsay Gardens, just north of the Castle Esplanade with its commanding views

Patrick Geddes / Paul Galeron / J.G. Bartholomew, *Suggested Plan for a National Institute of Geography* (1902)

of the New Town, where he constructed a romantic Scots-baronial and cottage-style set of flats that were run on a co-operative basis. At around the same time he acquired Short's Observatory, originally built in 1853 on the Castlehill, and converted it into a sociological laboratory and museum which he called the Outlook Tower. Along with the Granton Marine Station, Geddes used the Outlook Tower as the centre for his popular Edinburgh Summer Schools, which later took to the road, reaching Paris in 1900, where Geddes received further inspiration from geographer Élisée Reclus and globe-maker Paul Galeron. Reclus had created a globe 120 feet in circumference for the Paris Exposition of 1889, and had plans for one twice that size at the projected World Exposition in Paris in 1900. Galeron, who had created a planetarium with room for 100 spectators, proposed to dramatically improve the Outlook Tower, not only by including Reclus's two giant globes, but by visualising different geographic regions, from the local to the international, throughout its five storeys. The top storey housed a camera obscura for visualising Edinburgh and its interconnected human and physical landscape directly; the floor below included a large-scale model of the city and region; and below that was a room focused on Scotland and its history and geography. The floor below this was devoted to the British Empire and English-speaking countries, followed by the Europe Room, with the world as a whole on the ground floor. Rather than following the Art Nouveau style then popular on the Continent (but which never caught on in Scotland except in Glasgow), the building was to have been an eclectic grab-bag of styles ranging from Romanesque to Scottish Baronial and Italian Renaissance, but which nevertheless cohered into a dramatic whole.

The plan was printed by Bartholomew, and published in the *Scottish Geographical Magazine* of 1902, with a supporting note by Geddes, as well as this plea by John George Bartholomew: 'Our Institute of Geography would meet the wants of all ages and classes. It would supply much of what is needed to equip us for the practical work of life, and also what might satisfy and refine our leisure, enrich our human sympathies; it would be a temple of science in the truest sense.' For various reasons, including lack of funds, the Institute was never built, but Geddes's enthusiasms and influence continued unabated. In 1910 he was invited to create a Cities and Town Planning Exhibition in Edinburgh, as part of the first international town-planning conference organised by the Royal Institute of British Architects, and this work subsequently travelled to London, Dublin, Belfast and Ghent. After the First World War, Geddes spent an increasing amount of time abroad in the Empire and France, but continued to exert a powerful influence over town planning in Edinburgh and other parts of his native land.

RIGHT. The proposed Institute would house two giant globes, one terrestrial and the other celestial, echoing the work of Élisée Reclus.

OPPOSITE. The visualisation of the world on the ground floor would have been through a giant panorama, a terrestrial globe, and rooms devoted to particular countries, bringing together work from different disciplines and media.

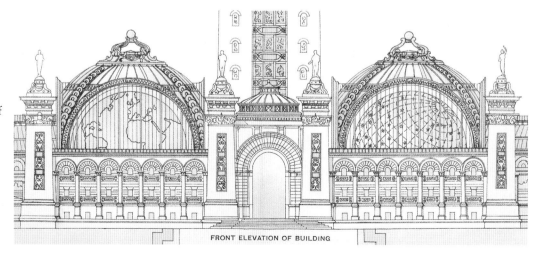

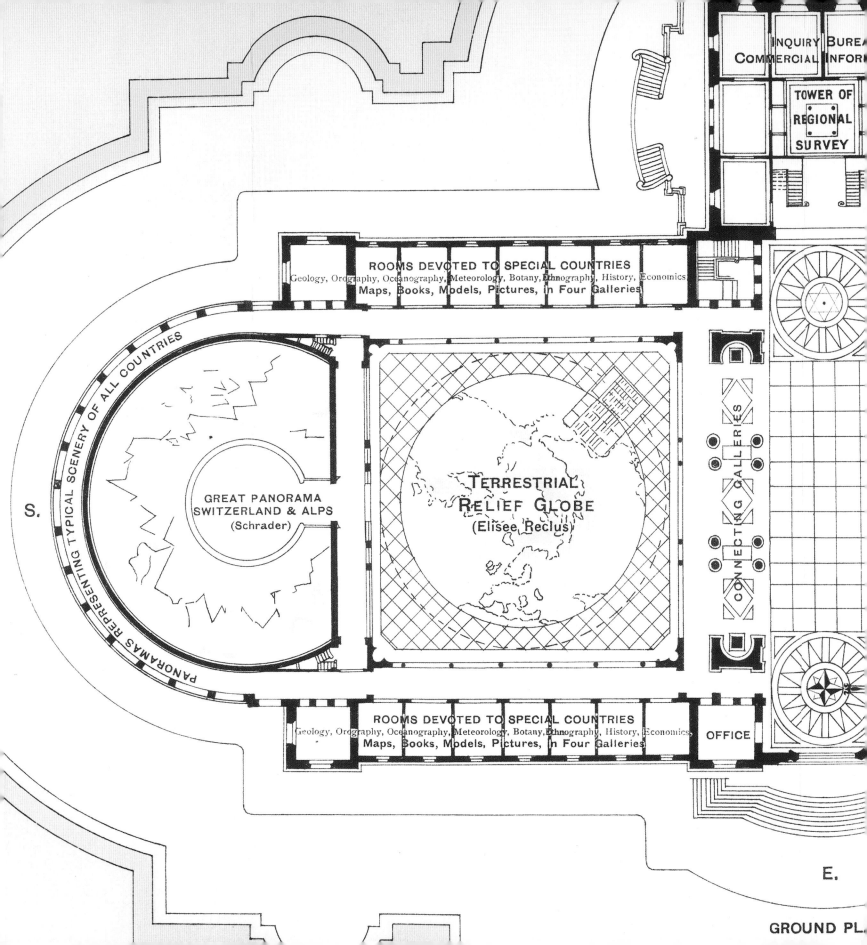

# 1903a

Banishing markets and slaughterhouses

Edinburgh's meat trades had a somewhat troubled history of being chased around the town over the centuries. James Gordon's flesh-stalls on the Royal Mile [1647a] were relocated to a shambles near the Nor' Loch in the 1680s [1742]. However, the construction of North Bridge allowed passers-by to see everything a little too closely, and 20-foot-high walls were constructed. Following demolition in the 1840s to make way for the railway, a new municipal slaughterhouse was constructed in the midst of high-density housing in Fountainbridge in 1850, but this soon faced sharp criticism from public health officials, including Dr Henry Littlejohn [1866], amid an increased awareness of meat-borne diseases, and an ever-decreasing public tolerance for butchery in a city centre that was already becoming an important tourist and retail-shopping destination.

The growth in volume of the inter-related cattle, horse and corn markets, historically centred on the Grassmarket, likewise caused problems by the early nineteenth century, especially as the pre-railway-age Grassmarket was also a centre for 'carriers', the group name then applied to inter-city and local messengers and parcel-delivery services. To reduce the congestion, a new cattle market was built at Lauriston Place. The arrival of the railways, however, only exacerbated the problem: former livestock markets outside Edinburgh – for example, the old ewe market at House o' Muir – were superseded as drovers began sending livestock directly into Edinburgh by rail. In 1902 alone, some 50,000 cattle and 250,000 sheep were driven through the streets of the city, most converging on the cattle market in Lauriston Place, with others going to the Haymarket or Valleyfield Street by the Tollcross. The construction of the new Central Fire Station directly adjacent to the cattle market in 1900 added to the chaos, as

---

Edinburgh Town Council / J.G. Bartholomew, *Plan of Edinburgh Showing the Areas Proposed to be Acquired and the Markets and Slaughterhouses Proposed to be Removed* (1903)

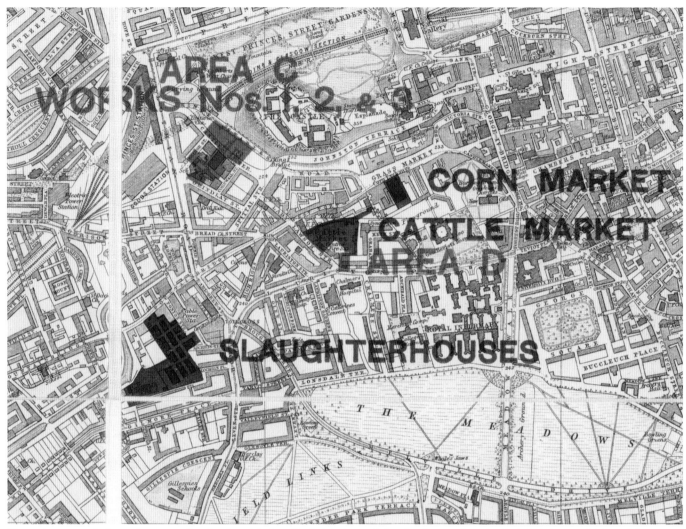

The main proposals focused on removing the Corn Market, Cattle Market and slaughterhouses, but also included the acquisition of the Synod Hall site on Castle Terrace (Area C), and a piece of land by the new Fire Station (Area D).

drovers fought it out with firemen and their horse-drawn fire engines. Clearly something had to be done, but the Town Council lacked suitable sites for relocation, and the powerful meat industry was in any case keen to retain its central sites.

All this would change in 1901, when the Trinity Hospital Commission sold the Town Council 160 acres of land centred on Gorgie Farm, thus commencing a long and complex process of moving the city's meat trades to this new centralised site. As expected, there was significant opposition from the industry, as well as a public enquiry, which delayed construction work. On the positive side, the sale of the valuable central sites defrayed costs, and the significant public benefits in terms of improved regulation and inspection were important arguments in the scheme's favour. Between 1907 and 1911 the new Corporation Slaughterhouse, the St Cuthbert's Association Slaughterhouse, a Market Restaurant and the Corn

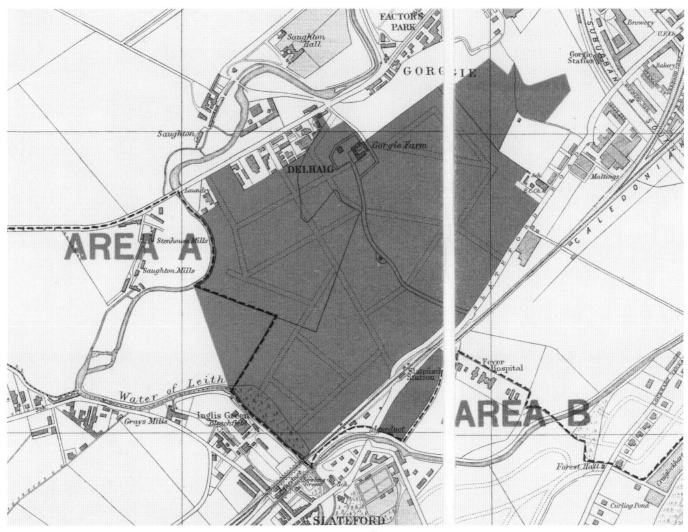

The proposed new Corn Market, Cattle Market, and slaughterhouses were all constructed by 1911 in the southwest corner of Area A in Gorgie.

Exchange were all built at Gorgie. A new street was laid out to connect Gorgie with Slateford Road, and was named after John Chesser, the convenor of the Markets Committee on the Town Council at the time, while the railway companies were eventually prevailed upon to construct local sidings.

Only 30 copies of this map, based on a standard Post Office Directory map, were printed by Bartholomew on 16 March 1903 for the Town Council's *Provisional Order regarding Markets, Slaughterhouses, Etc*. It usefully showed the location of the existing Corn Market, Cattle Market and slaughterhouses in blue, with the areas proposed for acquisition in red. As well as the large site for all three at Gorgie (Area A), it included the idea, then fashionable but soon to pass [1898b], that the new Usher Hall should be located on Castle Terrace on the site of the Synod Hall (Area C).

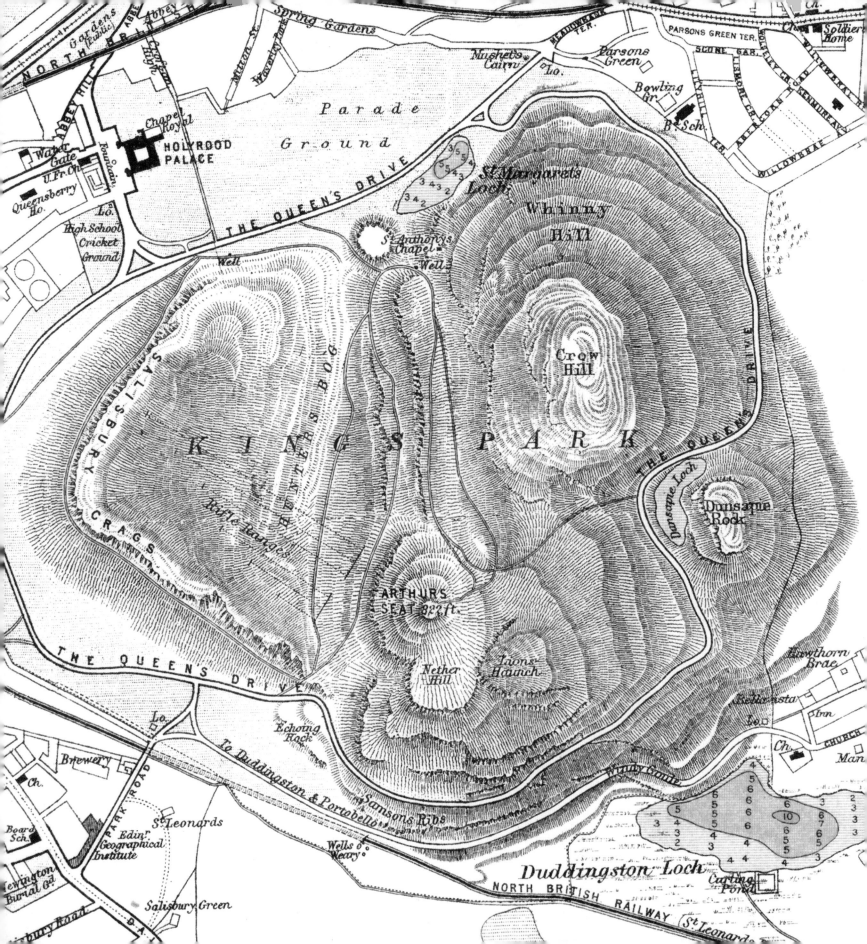

1903b

Triumph in the face of tragedy and governmental indifference: the Bathymetrical Survey

This map showing layer-coloured depths in Duddingston and St Margaret's Lochs was surveyed on 27 June 1903 by E.R. Watson and John Hewitt for the *Bathymetrical Survey of the Fresh-Water Lochs of Scotland*. This private initiative was based at the Challenger Office in Queen Street, and later in the Scottish Marine Station at Granton, located by the flooded quarry used to excavate stone for the harbour [1834b]. 'Bathymetry' literally means 'deep measure', and in practice refers to the underwater equivalent of topography: the mapping of undersea (or in this case under-lake) features such as peaks, troughs and ridges, initially via the primitive and time-consuming technique of depth-sounding using long poles, or ropes weighted with lead. The map usefully illustrates the important scientific and practical concerns that underlay comprehensive research on Scottish freshwater lochs at this time, not least because such features were roundly ignored by the official national mapping authorities.

The main impetus and driving force behind the *Survey* was oceanographer John Murray (1841–1914), who prepared the main scientific apparatus for the 80,000-mile HMS *Challenger* expedition of 1872–76, hailed in its own time as the most important voyage of discovery since the seventeenth century (Darwin and HMS *Beagle* notwithstanding). In addition to its foundational importance to the fledgling field of oceanography, the *Challenger* mission carried a floating natural history unit that catalogued more than 4,000 previously undiscovered species, and came under the overall direction of Sir Charles Wyville Thomson, Professor of Natural History at the University of Edinburgh. After Thomson's premature death in 1882, Murray co-ordinated the publication of the expedition results, and in doing so did much to transform oceanography from a disconnected set of

---

E.R. Watson and John Hewitt, *Duddingston and St Margaret's Lochs*, from *Bathymetrical Survey of the Fresh-Water Lochs of Scotland* (1897–1909)

different disciplines into a coherent and respectable branch of science in its own right.

Through his personal interest in yachting and first-hand observations of lochs, Murray realised that a systematic survey of bodies of fresh water would result in many new additions to scientific knowledge, and would be of great practical value to the growing number of engineers interested in supplying water to towns, as well as geologists and fishermen. He brought the subject before the Royal Societies of Edinburgh and of London, who made strong representations to the government on the value of such a survey in 1882–83, but without success: the Admiralty indicated that the work was not in the interests of navigation and so fell outwith their role, whilst the Ordnance Survey confirmed that their attentions were confined to dry land. The OS's reluctance may have had to do with the fact that the sizes and shapes of so many lochs were heavily weather-dependent, changing dramatically from season to season and year to year or even disappearing altogether during longer dry spells. In any case, as Murray concluded, 'there was no hope of the work being undertaken by any Government Department'.

From 1897, work began slowly on the lochs around the Forth, with some being surveyed two or three times with different sounding machines and methods before satisfactory results were obtained. Murray partnered with Fred Pullar, the son of an old friend, who constructed a sounding machine based on bicycle tubing and a drum carrying about 1,000 feet of galvanised wire. From a fixed position on the gunwale of a rowing boat, the device reeled the wire out into the water, and its three dials measured feet, tens of feet and hundreds of feet based on the rotating drum. (Previous systems had relied on manual counting of knots or painted marks.) Lake-bed deposits were also collected in brass tubes attached to the lead weight at the end of the sounding wire. Rather than attempt a time-consuming accurate locational fix for each sounding, the oarsman rowed in fixed lines across each loch using two poles as a transit, stopping at appropriate intervals, and evenly distributing soundings across the loch, thus equalising any errors. The device worked 'admirably and accurately' and was subsequently used by Robert Peary in his Arctic expedition of 1905.

Tragedy struck in February 1901, when Fred Pullar drowned at the age of 25 whilst bravely rescuing people who had fallen through ice on Airthrey Loch near Stirling. Murray considered abandoning the survey altogether, but at the request of Fred's father Laurence, the work continued – in fact, on a sounder footing, as he donated the sum of £10,000 to the project. Rapid progress was made: by 1906, some 60,000 soundings had been recorded on 563 lochs, with further biological and physical recording continuing until 1909.

A set of donated OS six-inch to the mile maps were traced onto cloth, on which the soundings were plotted and contour lines of depth drawn in at equal intervals. The areas within the consecutive contour lines were measured with a planimeter, and the volume of water and mean depths calculated. For Duddingston and St Margaret's Lochs, the maximum depths were 10 feet and 5 feet, respectively; the mean depths were 5 feet and 2½ feet; and the volumes of water were 4 million cubic feet and 500,000 cubic feet. Bartholomew then photographically reduced the tracings to half their original size and transferred them onto lithographic stones for colour printing.

The accuracy of the soundings have generally been confirmed by more advanced technology. For example, the Loch Ness bathymetric and seismic survey of 1992, using sonar, recorded a depth of 786 feet: a difference of only 4 per cent from the original bathymetrical survey completed in 1904. In a backhanded compliment, OS maps published after 1910 began to record the submarine contours of these and other lochs, copying this information directly from the *Bathymetrical Survey*.

PLATE CX

BATHYMETRICAL SURVEY OF THE FRESH-WATER LOCHS OF SCOTLAND
UNDER THE DIRECTION OF
Sir JOHN MURRAY, K.C.B., F.R.S., D.Sc., AND LAURENCE PULLAR, F.R.S.E.

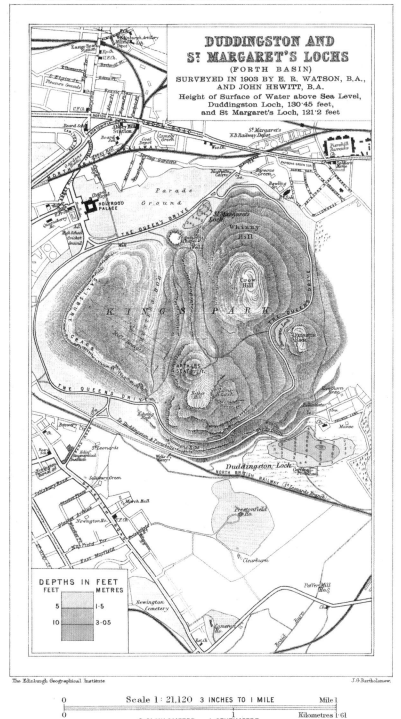

Published by the Royal Geographical Society 1905

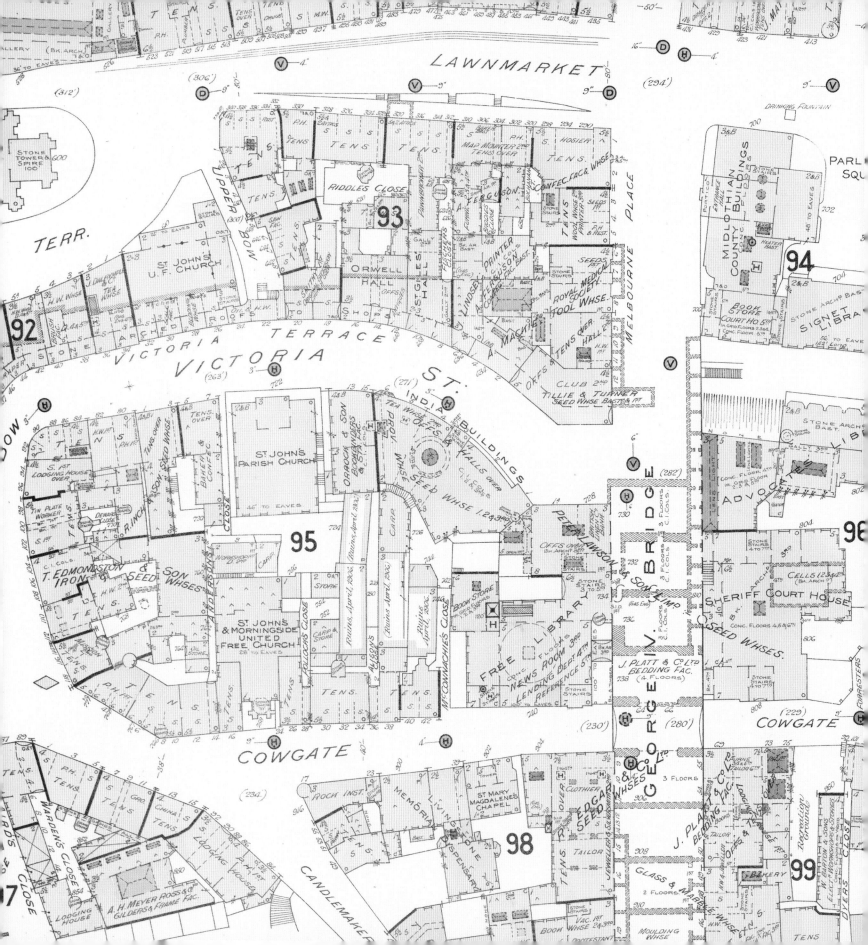

1906

A Goad fire-insurance plan of the George IV Bridge environs

Charles Goad (1848–1910) was born in Surrey, but moved to Canada in 1869, working initially as a civil engineer on various railway projects. In 1875 he spotted a demand for street maps that specifically showed information related to the risk of fire, and set up the Charles E. Goad Company in Montreal. Assisted by teams of surveyors, Goad mapped streets and buildings in new levels of detail, noting their construction materials, wall thickness, interior dimensions, room arrangement and function, type of roof, proximity to fire hydrants and fire-extinguishing appliances, and the locations of doors, windows and skylights. Building materials were colour-coded for flammability, so spatial concentrations of particular risks were immediately apparent. Initially, the business struggled and Goad continued his railway work, but soon it took off, and by the mid 1880s its series of large-scale insurance plans covered more than 1,300 places in Canada.

Far from being a Victorian or New World phenomenon, fire-insurance companies and fire brigades both originated in seventeenth-century England, with the latter owned and operated by the former. In the eighteenth century, buildings that were insured in Britain and Colonial America carried prominent 'insurance plaques' to indicate which insurance company's firemen were responsible for them in an emergency. Made of iron, lead or tin, these plaques often consist of a company logo above a five- or six-digit policy number, mounted above the front door. The Scottish system was not substantially different, even to the extent that at least one London-based insurer (Sun Fire) was selling policies and operating fire engines in Edinburgh by the late Georgian period. Despite fairly major fires in the Lawnmarket in 1725, Bishops Land in 1813, and West Bow in 1817, recommendations for the formation of a public fire service on military lines were ignored – that is, until after a fatal fire broke out in a spirit dealer's shop in Royal Bank Close in June 1824.

Charles Goad, *Fire Insurance Plan of Edinburgh*, sheet 12 (1906)

The interior of the Edinburgh Central Library, constructed 1887–90, with its News Room on the third floor, Lending Department on the fourth floor, and Reference Department on the fifth floor.

Edinburgh's 65-man, six-engine municipal fire service was formed in October of the same year, with the Town Council contributing 50 guineas per annum and two engines, and 21 different insurance companies £275 apiece. Unfortunately, the fledgling service utterly failed its first major test, the Great Fire of 15–17 November 1824 [1817].

In 1885, Goad returned to England and made London his base for a similarly comprehensive coverage of British towns and cities. Between 1888 and 1896 he produced plans for a number of Scottish towns, including Campbeltown, Dundee, Glasgow, Greenock and Paisley, as well as Edinburgh and Leith, with further updates after the turn of the century. He also expanded overseas, with plans produced for France, Denmark, Egypt, the Ottoman Empire, the West Indies, Venezuela, Chile and South Africa. After his death in 1910, his three sons continued the business, which survived as an independent company until 1974 and still produces insurance-risk plans today, as part of the Experian Group.

This detail of the George IV Bridge environs is from a Goad fire-insurance plan of 1906, and amply illustrates such plans' unique value then and now. Brick and stone buildings are shown in red and timber buildings in yellow; skylights are blue, as are hydrants (H) and valves (V) in the streets. A crucial type of information omitted from Ordnance Survey maps is the function of each building, including in this case tenements (TENS) and dwellings (D), shops (S), halls, churches, and a number of manufacturing premises including a smithy, a printer, a bookbinder, a confectionery factory, a bedding factory and a dispensary. All the buildings' internal layouts are shown, including features such as hoists (usually an H within a square box) and lifts. The map also gives details of the warehouses and storage spaces beneath George IV Bridge and their owners – also not included on OS maps. The attention to detail by the Goad surveyors is impressive, and simply gaining access to such a wide range of buildings and up the darkest vennels and wynds was an achievement in itself. Note, for example, the buildings marked as 'Ruins, April 1906' in the rear yards between Pollock's Close and Mr Connachie's Close on the north side of Cowgate.

At this time, the area depicted housed a large concentration of seed merchants and their warehouses, including Peter Lawson & Son, and Tillie & Turner at the top of Victoria Street; R. Inch & Son and T. Edmonston & Son on West Bow; and R. Edgar & Co. on George IV Bridge. As its name might suggest, the Grassmarket had been the Edinburgh area's main site for buying and selling corn, grain and seeds for centuries, and in 1849 a grand new Corn Exchange building had been constructed on its south side to a design by David Cousin.

Several of the public buildings shown survive today, including the India Buildings, the Free Library (opened in 1890), the Midlothian County Buildings (rebuilt the year before the map was produced), the Signet Library and the Advocates Library. The Sheriff Court House on the east side of George IV Bridge – whose every cell and window are helpfully shown for anyone planning an escape, for reasons of fire or otherwise – was demolished in 1937–38 to make way for the construction of the National Library of Scotland's George IV Bridge building.

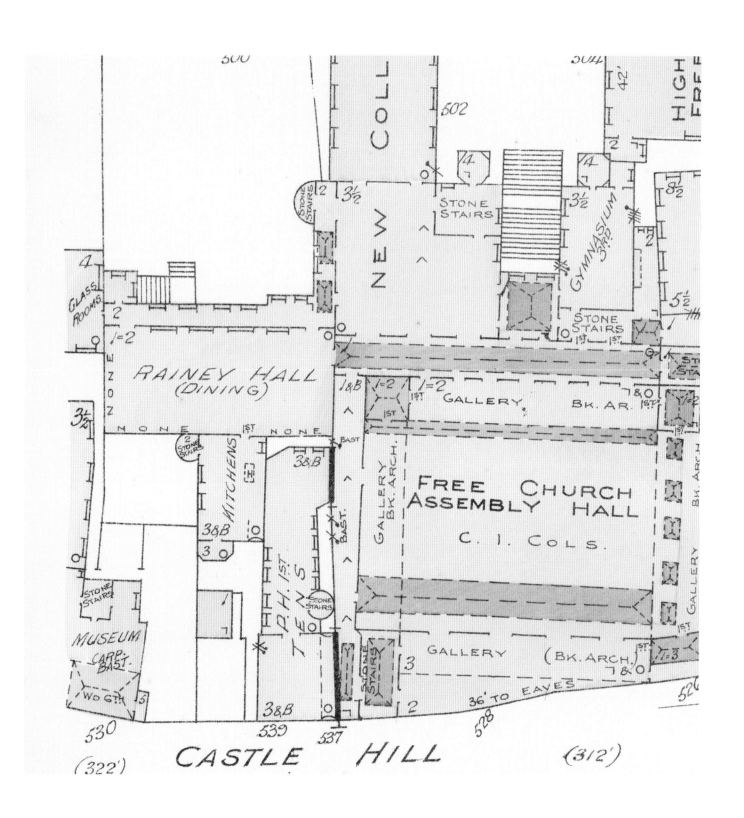

1919

A colour-coded chronology of Edinburgh's historical development

This striking and original map by John George Bartholomew illustrates several special qualities: his expertise and confidence in geography and cartography as a means of understanding the world; his interest in and affection for Edinburgh; his honorary secretaryship of the Royal Scottish Geographical Society and editorship of their *Scottish Geographical Magazine*; and his innovative use of colour lithography. Whilst recognising the commercial value of popular cartography, John George made particular efforts to produce new academic and scholarly maps throughout his life, often in collaboration with leading academics and researchers. His keen interest in furthering the discipline of geography led him to found the Royal Scottish Geographical Society in 1884, and to promote the founding of a chair in geography at Edinburgh University.

The map uses different colours to indicate the dates of the city's expansion: a red tint used for 'Old and Medieval Edinburgh', i.e. before 1750; blue for 'Renaissance', i.e. neo-classical Edinburgh, 1750–1850; and brown for 'Modern' Edinburgh, 1850 to 1919. Shades of these colours indicate shorter periods within these broad ones. Edinburgh's pattern of growth was of course much more intricate and interesting than a simple concentric expansion outwards, and the map captures this admirably: for instance, through the red colouring applied to a number of old roads, castles, tower-houses, mansions and place-names even where these had become submerged amid later development.

Bartholomew mapmakers were also masters of re-using the same base map, drawn on a copper plate, overlaid with different colour distributions to portray a wide range of different subjects. This base map at 3.5 inches to the mile or 1:18,100 had been used for the standard Post Office Directory maps for many years, and others including Bartholomew's

J.G. Bartholomew, *Chronological Map of Edinburgh Showing Expansion of the City from Earliest Days to the Present* (1919)

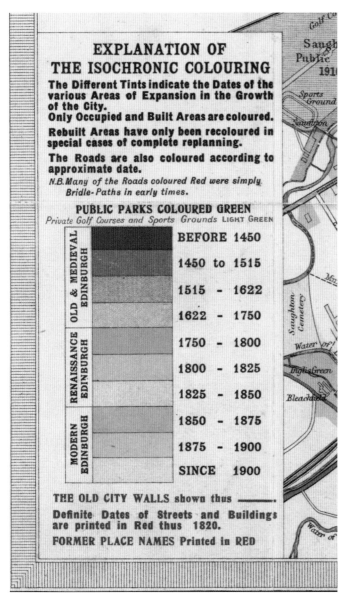

through his own knowledge and original research.

John George's summary of Edinburgh's cartographic history is also illustrative of the prevailing positivist assumptions behind cartography: that its history was simply one of progressive improvement from sketch maps to surveys, with ever-increasing exactitude. According to his *Summary of the Mapping of Edinburgh* table,

Before 1540	There were no published Maps of any kind
1540–1750	Only Sketch-Maps and Views of the City (Including Gordon of Rothiemay's fine plan, 1647)
1750–1855	Various Good Maps and Plans from Private Surveys
After 1855	The Ordnance Survey followed by Exact Maps

Although these views, in a diluted form, are still held to various extents today, research and writing on the history of cartography since the 1980s have taken a much broader and less deterministic view. John George's brief history of the city included here is equally redolent of an older mindset, visible well into the 1960s, that Scotland remained 'medieval' until the final defeat of Jacobitism in 1746, whereas in fact, as is now generally accepted, Scotland participated in the wider European Renaissance under Kings James IV and V in the early 1500s. This was the result of an intolerant and factually inaccurate assumption that there had been a 1:1:1 relationship between Jacobitism, Catholicism and 'medieval' forms of art and thought.

The map was included as a fold-out sheet with 'The Early Views and Maps of Edinburgh 1544–1852' in Vol. 35 of the *Scottish Geographical Magazine* (1919). This was a significant scholarly exercise in which Patrick Geddes [1895, 1902] wrote on the 'Beginnings of a Survey of Edinburgh'; architect Frank Mears, who had assisted Geddes on the Survey of Edinburgh for the Town Planning Exhibition of 1910, wrote a paper on 'Primitive' Edinburgh, accompanied by attractive conjectural sketch maps of Edinburgh c.1450; and William Cowan

own *Pocket Plan of Edinburgh* (1889), the *Survey Atlas of Scotland*'s Edinburgh town plans (1895 and 1912), and *Orographical plan of Edinburgh* (1902). Although we do not know for certain how John George drafted the map, he is known to have been a keen collector of historical maps of Edinburgh, and to have published several other maps portraying historical Edinburgh, which he supplemented

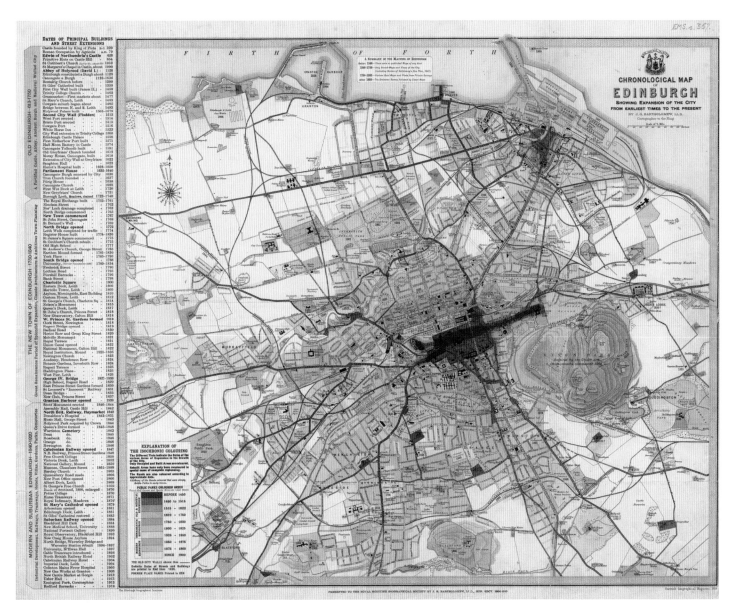

compiled one of the earliest bibliographies of Edinburgh maps. The editorial to this volume expressed the hope that the work of Bartholomew and Mears would prompt others to do the same for other cities in Scotland, but took a sombre rather than triumphalist tone, grappling with the realisation that knowledge and research had failed to prevent the catastrophe of the First World War.

This was probably one of the last maps drafted by John George Bartholomew. He had suffered since his early 20s from tuberculosis, which an eight-month voyage to Australia in 1880 had failed to clear; and in the spring of 1920, during another health-related trip to Estoril, Portugal, with his wife and daughters, he died.

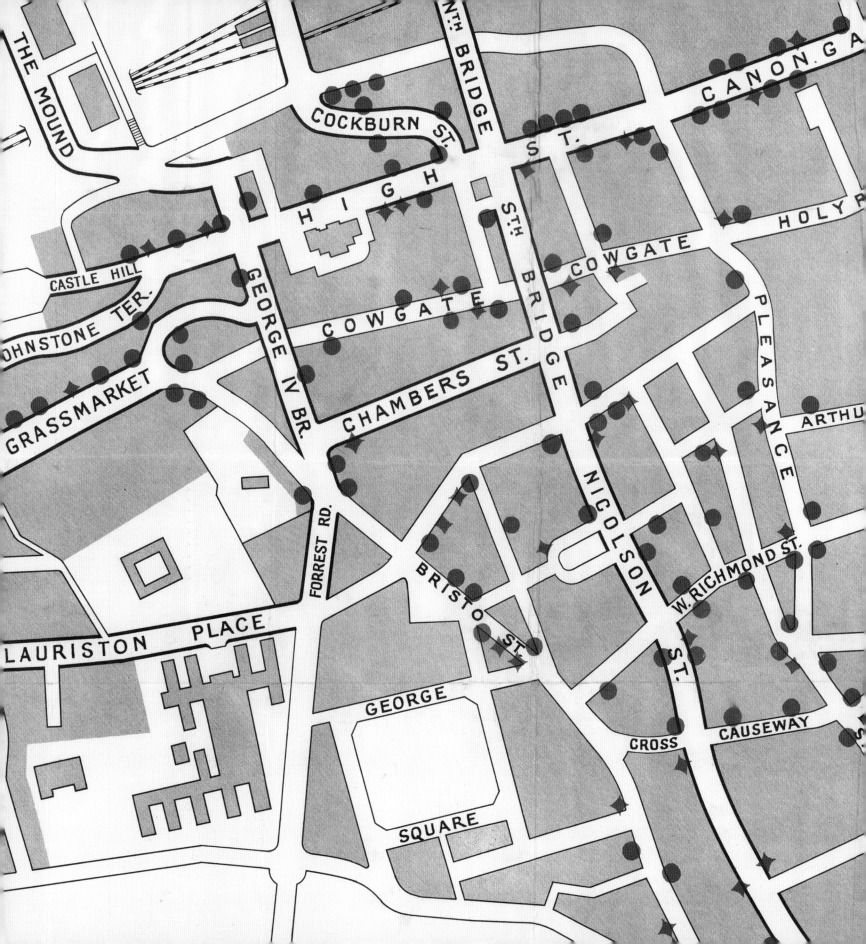

1923

Mapping an alcoholic 'bog of self-indulgence'

This map is a powerful example of cartographic propaganda, as well as illustrating the importance of the temperance movement at the height of its influence. Whilst all maps 'lie' in the sense that they select and distort the reality of the world, some do this far better than others; Bartholomew's clever selection of this particular part of Edinburgh, and their stark red symbolisation on a plain background of streets, creates a very striking graphic with a clear message.

During the nineteenth century, there was a steady growth in support for abstinence and for the total prohibition of alcohol consumption, internationally as well as within Britain. In large part this was ascribable to the growing worldwide popularity of Methodism: originally a sect of Anglicanism founded in the 1720s, whose members were strongly urged to abstain from alcohol entirely and to discourage its use by others. (Following the 1791 death of its founder John Wesley, a lifelong member of the Church of England, the movement came to be considered a church in its own right.) Support from religious societies such as the Band of Hope, founded in 1847 by a Baptist minister in Leeds, and the Methodism-based Salvation Army (1864), expanded temperance into a mass movement through alliance with left-wing groups and the campaigns for female suffrage. By 1890, a third of the 84 hotels listed in the *Edinburgh and Leith Post Office Directory* were temperance hotels, their role supported by temperance coffee houses, temperance halls and other public buildings. During the First World War, the consumption of alcohol was increasingly presented as unpatriotic, and this was coupled with higher taxes and tight restrictions on sales, paving the way for complete prohibition in the United States (1920–1933), Finland (1919–1932) and Norway (1919–1926), along with partial prohibition in Australia and New Zealand.

Under the Temperance (Scotland) Act of 1913, voters in local wards were allowed to hold a poll on whether their area

---

Edinburgh Citizens 'No Licence' Council / J. Bartholomew, *The Heart of Edinburgh* (1923)

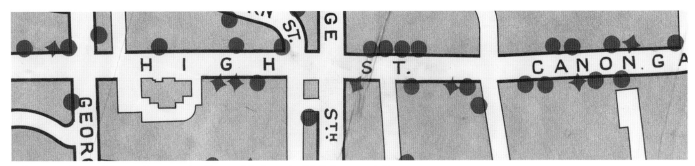
Pubs (as circles) and licensed grocers (as diamonds) along the High Street and Canongate.

remained 'wet' or went 'dry'. If more than 10 per cent of voters supported the latter, a formal poll was then held with three options: 'No Change'; a 25 per cent limitation in licences to sell alcohol; or 'No Licence', the abolition of all existing licences. The first polls in Scotland were held in November and December 1920, and although some districts did vote in favour of prohibition, particularly in the West of Scotland, over 60 per cent of votes nationally (and 67 per cent in Edinburgh) were for 'No Change'. However, this setback only encouraged temperance supporters to increase their publicity drive for the next opportunity at the polls in 1923, including sermons, newspaper articles, processions and demonstrations. The debate was heated but intelligent, correctly recognising that many people voting for 'No Change' *did* want the multiple problems associated with widespread alcohol abuse to be addressed, just not through limitation or prohibition. Unfortunately for the movement, reports of increased illegal drinking, drunkenness and crime from the US and Finland were already flowing in by 1923. But was this an inevitable response to prohibition, or just teething problems which increased enforcement would soon resolve? Duncan Maclennan, Chairman of the Edinburgh Citizens 'No Licence' Council, clearly took the latter view, and in a memorable speech to the National No-Licence Convention at the Music Hall in George Street, a month before polling in 1923, he announced that the people of Scotland would soon grow tired of seeing 'the wheels of the chariot of their deliverance stuck in the bog of their self-indulgence'.

The Edinburgh Citizens 'No Licence' Council ordered 500 copies of this very large poster, measuring some 30 × 40 inches, to be displayed at polling stations across Edinburgh on 1 December 1923. A map of the whole of Edinburgh would have been revealing – and Bartholomew had published such a map of Glasgow in 1884 for the Glasgow Young Men's Christian Association – but would have diluted the message so neatly delivered by this dense concentration in the Old Town and South Side. In the event, the support for 'No Change' in Edinburgh in 1923 grew slightly, to 68 per cent, with only 29 per cent in support of 'No Licence'. For prohibition to be implemented, the 'No Licence' option required a 55 per cent majority of a minimum 35 per cent turnout, so Edinburgh stayed 'wet'; and in spite of the temperance campaigners regrouping and contesting further polls in later years, support for temperance measures – both in Scotland and internationally – waned dramatically in the 1930s. After the Second World War, the global movement would largely be superseded in the public mind by campaigns against illegal drugs, with 'dry' jurisdictions limited to the most fervently religious areas. The Methodist Church in Britain had made total abstinence optional for its members by 1987, and now even allows ministers to drink responsibly in their own homes, but retains an institutional commitment to putting 'pressure on government and public opinion . . . to control consumption and reduce harm', whilst taking 'special care to avoid authoritarian attitudes which may be counter-productive'.

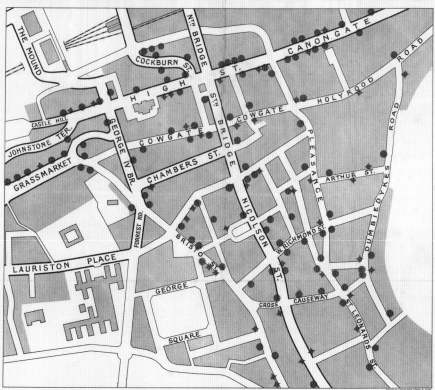

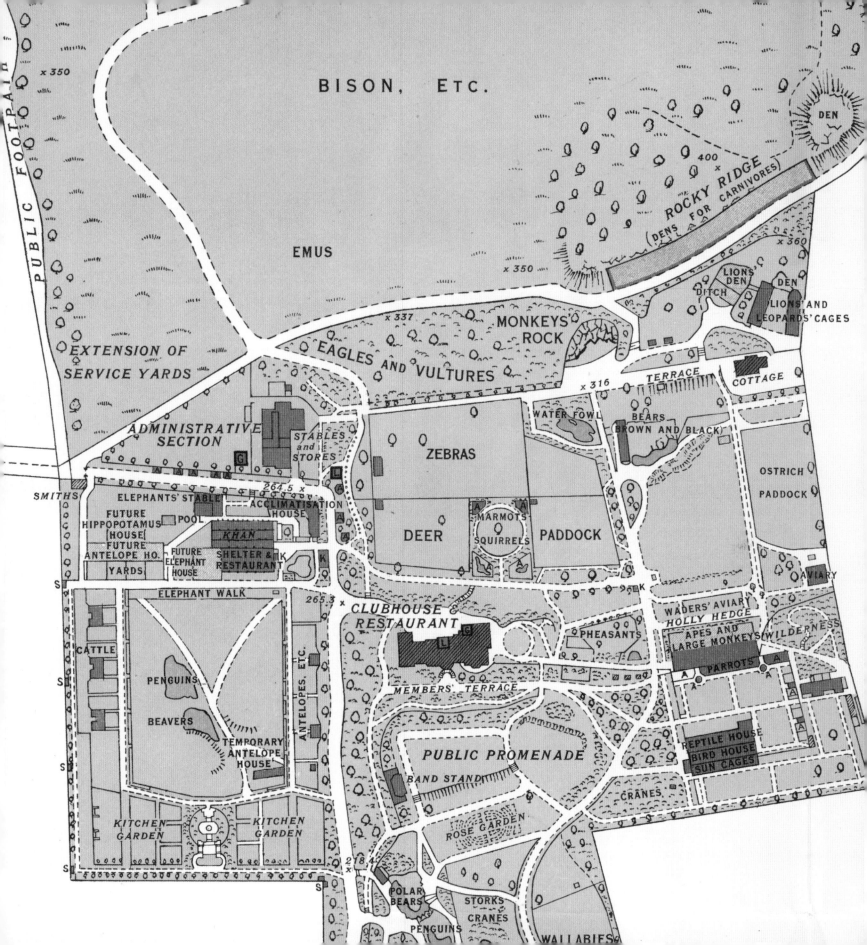

1932

The making of the Scottish Zoological Park

Encouraged by the opening of Regent's Park Zoo in London in 1828 and Dublin Zoo in 1831, the first Edinburgh Zoo was opened in 1840 at Broughton Park to the east of East Claremont Street. Its attractive landscaped grounds are shown on mid-nineteenth-century maps as the Royal Zoological Gardens, but the institution experienced financial difficulties and closed by 1858. Half a century would pass before the Zoological Society of Scotland formulated plans for a new zoo, along scientific and educational lines as opposed to the original 'menagerie' principles.

The new zoo was formally constituted as a registered charity by Thomas Hailing Gillespie in 1909. Gillespie was an Edinburgh solicitor, but with strong interests in zoology, and was a driving force in the field, acting as director-secretary of the Society from 1913 until his retirement in 1950. He also became well-known as 'the Zoo Man' on the Scottish Children's Hour radio programme in the 1920s, giving talks on animals and answering general nature questions. He campaigned for a Zoological Park in Edinburgh modelled on the more open, barless enclosures that had been used so successfully by Carl Hagenbeck's Stellingen Zoo near Hamburg.

The 75-acre site shown here was acquired from the estate of Corstorphine Hill House in 1913, rising on the southern slopes of Corstorphine Hill, with magnificent views to the east and south. The bounds of the area, including the original house from the 1790s and some of the paths and patterns of woodland, can be traced on the 1826 *Plan of the Estates of Ravelston and Corstorphinhill* [1826a]. The park was laid out by Patrick Geddes and his son-in-law Frank Mears [1919], who cleverly converted the quarry-pits from which stone for the zoo roads had been extracted into pools or sunken habitats for animals. Deeside-born Geddes was a strong advocate of regional study and the 'valley section', a model of a region

---

John Bartholomew & Son Ltd, *Scottish Zoological Park* (1932)

based on a river travelling from its source to the sea, and this provided a philosophical basis for the practical design of the zoo site. In Geddes's view, the concept of regions offered an antidote to nations and conflict, and provided a means for understanding different ecological zones; how man's occupations were supported by physical geographies; and how people could understand their place within particular environments and the world.

Only the southern area of the zoo was developed at first, up to the Administrative Section and Lions' Den, with the area further north retained as a golf course. The plan here shows the layout after the redevelopment of the frontage area, following the widening of the Corstorphine Road in the 1920s and the construction of the new Lodge and Carnegie Aquarium, opened in 1927.

Lord Salvesen became the Society's first president, and the Leith-based firm of Christian Salvesen generously donated regular consignments of penguins, particularly from South Georgia, when its whaling ships returned from Antarctica. As a result, Edinburgh Zoo was the first in the world to house penguins, as well as the first to breed them, beginning in 1919. The famous Penguin Parade, a voluntary perambulation of the penguins outside their enclosure, began by accident in the 1950s and still continues today. Another accidental zoo spectacle followed the escape of a colony of night herons from their aviary. Attempts to recapture them proved fruitless, but as the birds had few natural food sources in nearby rivers, they continued to roost nearby, and descended to devour fish at the sea-lions' feeding time. As both sexes of bird had escaped, they bred successfully for several years, numbering some 50 pairs by the 1960s, with an extended nesting range.

Bartholomew printed 10,000 copies of this colourful plan for inclusion in Gillespie's *Popular Official Guide to the Scottish Zoological Park*. It provides excellent detail of the layout and animals there in the early 1930s, and a wealth of supporting topographic information, including spot heights above Ordnance Datum, woodland cover, escarpments, and nearby footpaths and roads. Boosted by the arrival of giant pandas in 2011, the zoo now receives over 600,000 visitors a year, making it Scotland's second most popular paid-for tourist attraction after Edinburgh Castle. Contemporary plans of the zoo resemble this one in layout, but not in style, and lack something of the original vision and sentiment still visible in the 1932 plan. Much greater emphasis is now placed on instant information for a more international audience: pictograms of animals; symbols for facilities, event times and feeding times; and suggested routes, varying according to weather conditions and the visitor's level of fitness. The two 'Kiosks for the sale of Nuts, Postcards, etc' from 1932 have been substantially upgraded, and there has been a significant shift towards more exotic wildlife.

The Society was incorporated by royal charter in 1913 and granted the privilege of adding a 'Royal' prefix to their name in 1948, following a visit by King George VI. Edinburgh remains the only UK zoo with a royal charter. Christian Salvesen exited the whaling business in 1963 to become a general transport and logistics company.

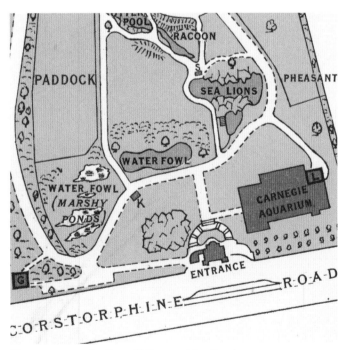

The entrance to the Zoo off Corstorphine Road, with its Lodge and Aquarium.

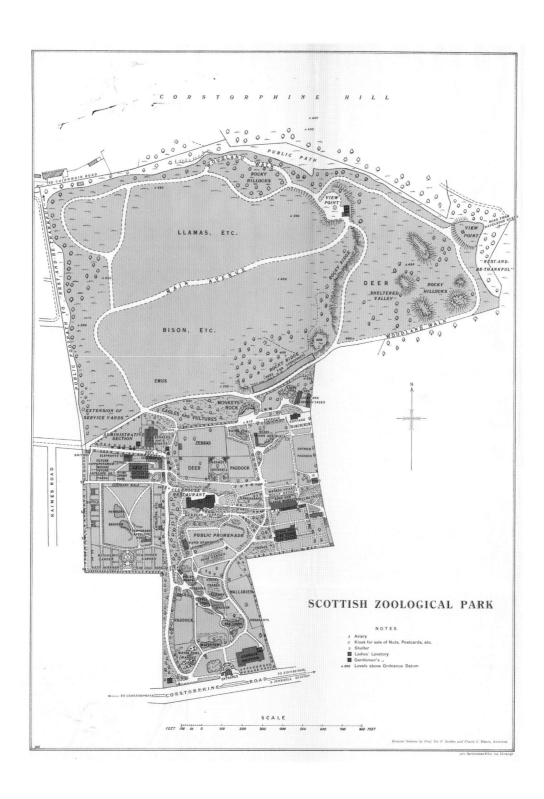

1934

Post Office Directory maps and urban history

Post Office Directories and their maps are rightly regarded as indispensable primary sources for urban history from the late eighteenth through to the mid twentieth centuries. Peter Williamson, a printer and publisher based in the Lawnmarket and Luckenbooths who also published the *Caledonian Mercury*, brought out the first *Directory for the City of Edinburgh, Canongate, Leith and Suburbs* in 1773, continuing with annual or bi-annual updates into the 1790s. Williamson was one of Edinburgh's most extraordinary characters of his time, or any other. Kidnapped as a young boy and sold into servitude in the Colony of Pennsylvania for £16, he worked out his indenture but was captured by a war-band of Native Americans. He escaped and secured a colonial army commission in Massachusetts in 1755, only to be captured again, this time by the French. Despite having had no formal education beyond the age of eight, he wrote a bestselling book based on his experiences in the New World, which incidentally brought down the ring of Aberdeen child-slavers who had sent him forth from Scotland in the first place. Newly wealthy from a combination of book royalties and legal damages won from the kidnappers, Williamson founded a private postal service with 18 fixed offices and four uniformed messengers, which carried letters and parcels weighing up to 3 lb within a one-mile radius of the Edinburgh Mercat Cross, as well as between Edinburgh and Leith. This service operated for two decades before being purchased *en bloc* by the General Post Office (GPO). Some time after his arrival in Edinburgh, Williamson affected the costume of the Lenape (Delaware) tribe, apparently wearing no other garb for the rest of his life; he was even buried in it.

Under the name *Post Office Edinburgh and Leith Directory*, various other publishers and printers continued

---

John Bartholomew & Son, *Post Office Directory Map of Edinburgh and Leith* (proof and finished copies) (1934).
Proof map © 1934. Published by permission of HarperCollins publishers

The Edinburgh printer and publisher Peter Williamson (d.1799), sporting his traditional costume of the Lenape (Delaware) tribe.

this work from 1805, usually as annual editions; Morrison & Gibb, based at Tanfield, took over the title in 1880 and published it until the 1950s. In its mature forms from the 1830s onward, it included alphabetical lists of personal names and addresses; a list ordered by street and house number; and a commercial trade listing, often with useful supplementary information and advertisements. Although originally the address information was compiled by Post Office letter carriers, from the 1840s the Directory companies were forced to compile and publish the volumes as purely commercial concerns, after questions were raised in the House of Commons about the improper commercialisation of Post Office materials and employees.

Street maps were increasingly regarded as an essential ingredient in the Directories as time went by. In contrast to many other maps, which were so often produced to promote future developments, these had to be scrupulously accurate with regard to what buildings and streets actually existed. The Directory companies funded the compilation of revised new maps on an annual basis from the 1820s, a process that would include several of the leading map engravers and publishers in Edinburgh: Lizars, Brown, Wood, Johnston and Bartholomew. From the 1850s, commercial mapmakers had excellent Ordnance Survey maps to use as a base, but nevertheless maintained their own street directory maps, updating these every year. Whereas the Ordnance Survey only revised maps of Edinburgh at intervals of 10 to 30 years, the Post Office maps provided more specific road and building updates every year, and (as we have seen) were often re-used by the Town Council, companies, and other entities for illustrating a variety of themes [1869, 1879, 1892a, 1903a].

Although for a time they lost the contract to W. & A.K. Johnston, Bartholomew published the Edinburgh Post Office map from the 1860s to the 1920s. The scale remained the same at 6 inches to the mile (1:10,560), but as Edinburgh expanded, the size of the map quadrupled, from 20 × 24 inches to 40 × 50 inches. Print runs were usually 3,000–4,000 per year, and with only marginal additions to be made, this work represented a steady and reliable income stream.

The Bartholomew archive also provides a useful practical insight into the map-compilation process. Bartholomew were desk-based cartographers, and sent a form letter each year to all the main institutions who oversaw new building work – the City Architect's Office, the Public Works Office and the Burgh Engineer's Office – as well as to institutions to which they knew changes had been reported, such as the Royal Infirmary and Edinburgh University. They would enclose copies of the old Post Office Directory map and request a returned copy with annotations to show change. These were then compiled at Bartholomew on a proof map (p. 260), usually with red to show additions and blue or black for deletions. This was then transferred to the main lithographic master for printing the new map (right). This particular detail shows the new buildings and streets being constructed at Granton and Pilton in 1934, providing a valuable record of construction work in progress.

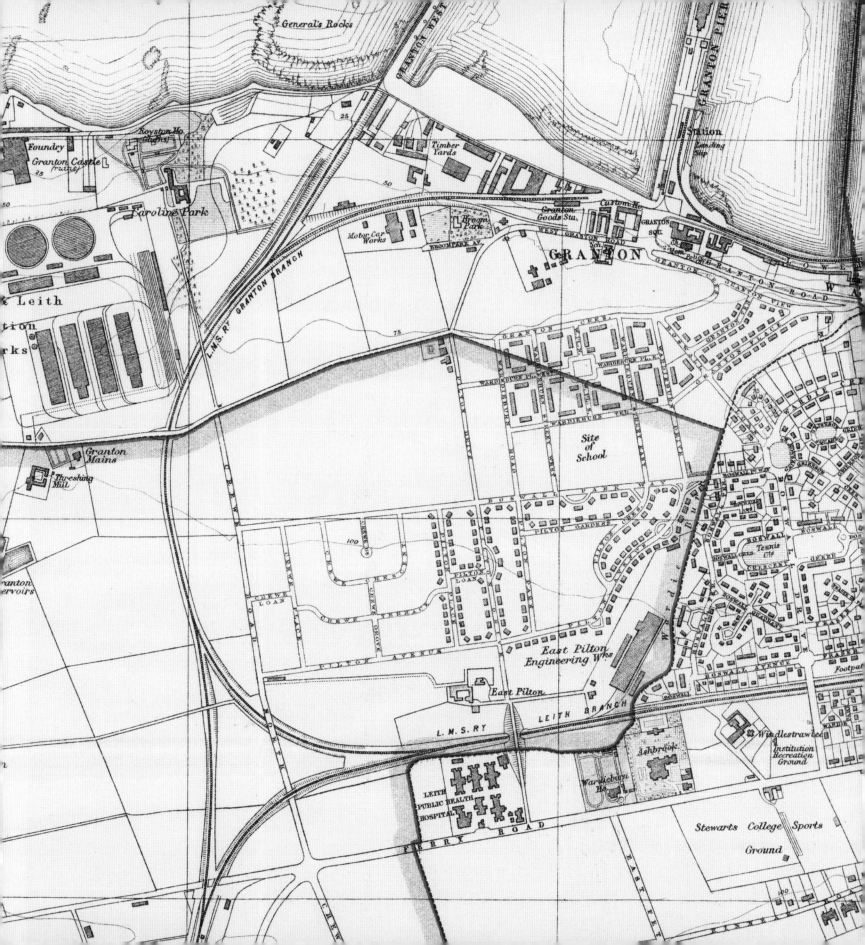

1941

A German bombing map of Edinburgh

Cartography has been closely associated with warfare from the earliest times, but as military technology has changed, maps have changed too in their function and content. Aerial bombardment with the specific aim of killing and demoralising civilian populations was a new phenomenon of the First World War and a predominant strategy of the Second, and both sides created maps in huge numbers to assist with these deadly campaigns.

This German military map of 1941 is overprinted with colour and symbols to pick out bombing targets in and around the city. These include railway stations, bus depots, hospitals, bridges and a range of industrial targets, as well as the Old College and National Museum on Chambers Street (with a symbol denoting 'art collection / cultural monument'). The West End and Gorgie were home to the greatest concentration of targets, including the power station in Dewar Place, a sawmill by Haymarket, an ironworks, a distillery, a brewery and a biscuit factory, all of which were highlighted with special symbols. The accompanying *Eintragungen* or listing on the reverse of the map numbered particular buildings.

Under wartime conditions, the easiest way of assembling detailed mapping of large geographic areas was by capturing enemy mapping and overprinting it with relevant military information. Here, a standard prewar Ordnance Survey 6 inch to the mile map (1:10,560) has been reduced to 1:15,000. Some of its target information was already present on the original map, and was simply translated from English into German; but much of it was supplied by other forms of reconnaissance including espionage, rectified aerial photography conducted in May 1941, and the use of published directories [1934].

Edinburgh did not, in fact, suffer major damage during the Second World War, especially compared with Clydebank, Greenock, or even some isolated ports like Fraserburgh; the

---

Generalstab des Heeres. *Stadtplan von Edinburgh mit Mil.-Geo. Eintragungen* (1941)

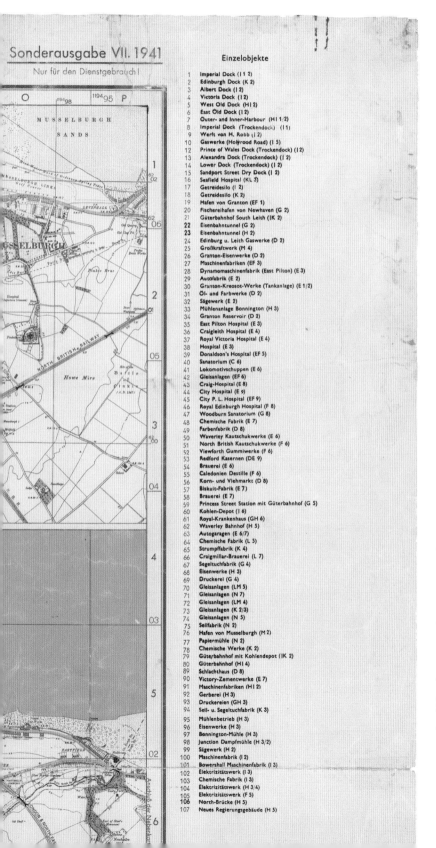

Luftwaffe paid far greater attention to the Forth Bridge and the Rosyth naval dockyard than to the Scottish capital itself. The first Luftwaffe attack on the bridge in October 1939 took everyone by surprise, despite it having been a well-known target of Zeppelin raids in the previous war. Repeated attacks on Scotland's eastern coast rapidly changed perceptions. After the German conquest of Norway in April–June 1940, it was widely believed that an invasion of Britain might be launched from that quarter via northeast Scotland – perhaps using glider-portable tanks that had been captured in Poland. The British Army hastily constructed a number of north-facing 'stop lines' consisting of hexagonal masonry pillboxes, ditch-and-bank anti-tank barriers, concrete anti-tank cubes, slit trenches, weapons pits, barbed wire, roadblocks made from old railway rails, and other obstacles, notably along the line of the Cowie Water by Stonehaven; and engineers stood in constant readiness to blow bridges that the German invaders might use as they headed south toward the Central Belt. Nevertheless, the number of troops available to defend this system in the aftermath of the fall of France was hopelessly inadequate, with a single division of about 10,000 men tasked with defending the entire east coast from Easter Ross to Grangemouth, compared with a German force in Norway numbering up to a third of a million.

The invasion of course never materialised, but between 1940 and 1942 there were at least 14 recorded bomb attacks on Edinburgh, which killed 18 people and injured more than 200. Many of the bombs fell on Leith and Granton, but a number of other areas were hit, including Craigentinny, Portobello and Seafield Road, Abbeyhill, Craigmillar, Niddry, Corstorphine and Gorgie. One 500-lb bomb fell in Gorgie on 27 September 1941; although the raiders may have been aiming for the distillery, they in fact hit a bonded store at the

This numbered listing of 'Einzelobjekte' (individual features) pinpoints docks, hospitals, schools, factories, gas works, stations, breweries, tunnels, reservoirs, and other locations for possible bombing.

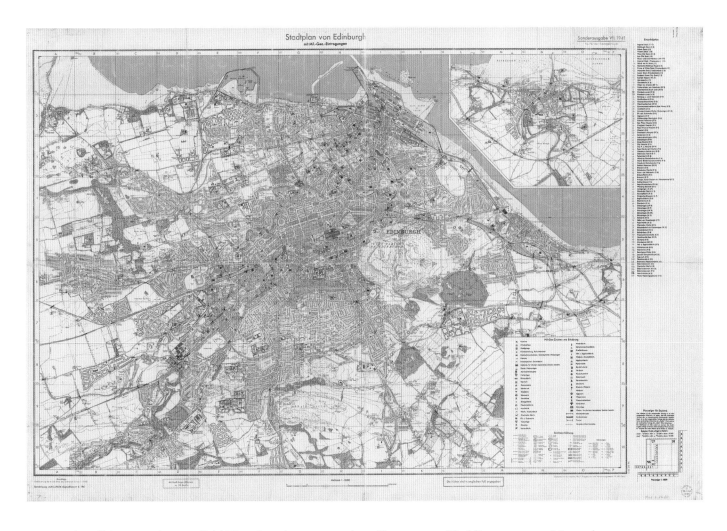

corner of Duff Street and Downfield Place housing many oak casks of whisky, which exploded and sent spirits flowing down the street. In addition to a wide range of anti-aircraft artillery and the top-secret radar installations co-ordinated from Dollarbeg Castle in Clackmannanshire, the air defences of the city included 602 (City of Glasgow) Squadron, based at Drem in East Lothian, and 603 (City of Edinburgh) Squadron, at Turnhouse. Both squadrons were equipped with state-of-the-art Spitfire fighters, and their pilots were credited with two notable victories: the first enemy aircraft shot down in British waters during the war – a Ju-88 off Musselburgh on 16 October 1939 – and the first shot down on British soil, an He-111 near Haddington on 29 November.

Although these are probably underestimates, official figures state that across Scotland 2,298 people died, 2,167 were seriously injured and 3,558 slightly injured as the result of enemy bombing. To put this into perspective, Allied bombing raids on Hamburg alone in July 1943 caused some 45,000 deaths, mainly because the tactics and aircraft were different – but both sides relied on detailed mapping. This map provides a chilling reminder of the dangers of life in wartime Edinburgh, as well as the vital role of enemy mapping for both sides, whether captured in the field or purchased by 'tourists' in the prewar period.

1946

RAF photography and Operation Revue

Although the First World War had been the major stimulus for the use of aerial photography in military reconnaissance and surveying, Britain's Ordnance Survey did not generally employ aerial survey methods in the interwar period. During the Second World War, however, millions of air photos were taken by the Royal Air Force, and their value for both military reconnaissance and topographic surveying was proved beyond all doubt. In 1945 the Air Photo Division was established within the Ordnance Survey, utilising newly redundant aircraft and personnel from the RAF. The Air Ministry, meanwhile, agreed that the RAF would undertake Operation Revue: an aerial photographic survey of Great Britain that was later extended to cover large parts of Europe.

Around 500 sorties were flown in Scotland, resulting in the collection of over 280,000 photographs. Mosaics like the one shown here were produced from these by the Ordnance Survey, mainly as a quick and cheap expedient before proper ground-based surveying could be carried out; but they were also intended to benefit government departments dealing with town planning and transport. Creating the mosaics was itself a technically skilled job, involving rectifying the original photographs to compensate for distortion, carefully cutting overlapping sections and pasting them together to form a composite image, then adding lettering for major towns and straight borders. The air photo mosaics of Scotland were all published at 1:10,560, though some urban areas of England appeared at the larger scale of 1:1,250. Operation Revue was initially scheduled for completion by the end of 1947, but many areas were re-flown in 1948–49 to correct problems caused by issues including cloud cover and camera failure. Cameras that were heated to prevent lens-fogging had already been created by Sidney Cotton, an Australian spy who had flown a number of successful aerial reconnaissance missions over prewar Nazi Germany.

Ordnance Survey, *Air Photo Mosaic 36/27 SE* (1946)

Owing to postwar budgetary constraints, high-speed wartime aeroplanes such as Spitfires and Mosquitos had to be used, rather than the slower Ansons that were in fact more suited to peacetime aerial survey. To allow stereoscopic capture, the aircraft were specially modified to take two cameras – one beneath each wing; this technique had been pioneered during the war and its '3-D' effect was especially useful for distinguishing ground features from one another based on height. Though lenses up to 40 inches in diameter were used by the wartime RAF, most were 8.5 × 7 inches or smaller. For reconnaissance pilots, most of whom flew solo, finding the correct balance of air speed and photographic exposure time required enormous skill – even when they were not being targeted by enemy air or ground weapons.

This aerial photo shows the Canongate and Holyrood, with the high buildings fronting the Royal Mile casting a significant shadow on the south side. In the war, it was considered preferable to take photos 1–2 hours before or after 'solar noon', so that shadows would be available to aid in interpretation, but not be so long that they obscured neighbouring objects. The light-coloured rectangular structure is Moray House, but the greater proportion of large buildings around Holyrood were still breweries [1813], including St Ann's to the northeast, the Abbey Brewery to the west (where the new Parliament now stands), and Park Ale Stores further south by the two large gas cylinders of the Edinburgh Gas Light Company. Bell's Brewery and the Pleasance Brewery can be seen to the east of the Pleasance at the bottom.

George IV Bridge to George Square, with the dome of the McEwen Hall in the centre.

Holyrood Abbey and Palace, with the Abbey Brewery at the bottom of Canongate to the left.

St Andrew's House on Regent Road.

St James Square and Leith Street.

Moray House, at the junction of St John Street and Holyrood Road.

Gas holders and the Park Ale Stores, south of Holyrood Road.

1949

Postwar visions for the motorcar age

As shown by the work of Craig [1767], Kirkwood [1817], Stevenson [1819b] and Cousin and Lessels [1866], Edinburgh Council have had a long history of using maps for urban planning. This trend became even more prominent in the twentieth century, and one of the most famous and cartographically most impressive urban-planning maps was created in the late 1940s by the celebrated town planner Sir Patrick Abercrombie (1879–1957), who was born in England and educated in England and Switzerland. The council invited Abercrombie to Edinburgh to collaborate with planning officer Derek Plumstead in 1945, hard on the heels of the former's widely acclaimed postwar reconstruction plans for London and its environs (1943–44). Concurrently with an academic career in Liverpool and London, Abercrombie was increasingly sought after, and created a prize-winning new plan for Dublin in 1914, followed by the re-planning of a number of English towns including Doncaster (1920), Sheffield (1924) and Bristol/Bath (1930). Second World War bomb devastation – notably in Plymouth, Hull and Bournemouth – provided him with blank canvases for quite radical schemes, enhancing his reputation still further.

Although Edinburgh had largely escaped bomb damage [1941], some felt that its planning problems were similar to those of the bombed towns. Buoyed also by the postwar surge in enthusiasm for public ownership and management, Abercrombie and Plumstead saw themselves very much as professional experts, and cities as requiring 'top-down' order that it was the planner's mission to enforce, engineering them into more efficient forms through new technologies, particularly the automobile. This authoritarian and technocratic

---

Patrick Abercrombie and Derek Plumstead, *Plate XXIII City Centre Plan* and *Plate XXXV Scale Model of Princes Street Scheme*, from *A Civic Survey & Plan for the City & Royal Burgh of Edinburgh* (1949). By permission of the City of Edinburgh Council

outlook was all the more surprising, given that Abercrombie saw himself as a follower of the relatively minimalist and 'grass-roots'-based planner Sir Patrick Geddes – further proof, perhaps, that any ideology can be transformed into its opposite in the space of a single generation, often without its name being changed. After Geddes's death in 1932, Abercrombie was increasingly influenced by Swiss-French modernist Le Corbusier and by road-traffic expert Sir Herbert Alker Tripp. In the 1940s, Abercrombie continued to advocate Geddes-style detailed civic surveys [1902] and the socio-cultural independence of the city-region; but by and large he had moved on, into a brave new world that would soon be dominated by younger 'brutalist' architects like Sir Basil Spence.

This map illustrates some of Abercrombie and Plumstead's ideas for central Edinburgh, including an inner ring-road scheme incorporating a three-decker Princes Street, offering an 'enchanting view for the visiting motorist and a wonderful opportunity for a feat of engineering and of architectural achievement'. The top storey was intended primarily as a service link road, the middle one for car parking and a promenade, and the lowest for through traffic with its southern side facing Princes Street Gardens planned as a continuous colonnade. Abercrombie firmly believed that 'a system of suitably designed dual carriageway roads is necessary and will become more so in the future as petrol and other restrictions are removed'. The Princes Street dual carriageway continued westward under the Caledonian Station to emerge at the west side of Rutland Square, from whence a large roundabout carried the ring-road south along an expanded Melville Drive forming 'a tree-lined boulevard flanked by the sweeping Meadows'. Here it met another large roundabout near Hope Park, with a dual carriageway bulldozing on north through the Pleasance to bypass the bridges, proceeding over the Cowgate but under the High Street, and then crossing the Waverley Valley as a flyover to complete the circuit at the eastern end of Princes Street. Another roundabout and flyover by Jeffrey Street (illustrated best in the model) curved on stilts down to the Mound.

This was merely one component. To improve communications further, and relieve congestion at Waverley, Abercrombie and Plumstead planned a new inter-city railway station at Morrison Street, replacing the Caledonian, while Waverley's focus would be changed to suburban traffic. A new railway tunnel beneath the Meadows (shown with a dashed line) would connect with the London line at Duddingston, especially to cope with an anticipated expansion in coal traffic from the Lothians. The colouring on the map also illustrates the zoning principle behind Abercrombie's thinking, and the influence of Tripp's 'precinct theory'. Edinburgh was divided into a set of zones – industrial, commercial, educational, residential and recreational – that were to be safeguarded from outside influence: traffic on communication arteries would be encouraged to reach its destination as fast as possible, without invading the precincts. Under this theory, central Leith and Gorgie were to be redeveloped as industrial zones with their populations largely removed; Portobello developed as a seaside suburb 'reminiscent of Brighton'; and other mixed and messy zones on the ground steadily harmonised into one or another predominant type.

Though all too easy to caricature for its arrogance and destructiveness, especially with the benefit of hindsight, the plan itself is an impressive work, filled with optimism, vision and a conscientious sincerity. Abercrombie and Plumstead presented their advisory plan for the Town Council in October 1947, and two years later published the full 111-page report. Copiously illustrated with more than 80 plates, photographs of models, views and sketches, its aesthetics and vigour continue to impress readers today. A large team compiled very detailed geographical and statistical information, partly presented as 25 hand-sketched thematic overlays on top of a set of specially prepared base maps: population distribution and densities, family sizes, dwelling conditions and ownership, historic buildings, open spaces, road traffic and even road accidents. Although the plan's central and most visible proposals were obviously never adopted, it continued to influence much postwar planning, and several of its recommendations for more peripheral areas – notably the outer radial of the City of Edinburgh Bypass – were implemented by the 1960s.

1983

Mapping for a Soviet tank advance on Edinburgh

Given that the earliest detailed maps of Edinburgh were made by foreign military engineers who accompanied the Earl of Hertford in his 'Rough Wooing' of Scotland, it is perhaps fitting that some of the most recent really detailed maps have also been made by military cartographers – this time, in the Soviet Union.

This map of Edinburgh by the Soviet Army was produced in 1983, at a particularly tense phase of the Cold War. The Soviets had invaded Afghanistan in 1979, and in 1983 shot down a Korean Air Lines flight that violated Soviet airspace. Both Margaret Thatcher, as British Prime Minister from 1979, and Ronald Reagan, US President from 1980, denounced the Soviet Union and its ideology, and significantly increased military spending, vowing to check the ambitions of the 'evil empire' around the world. On the night of 26 September 1983, nuclear war nearly broke out when Soviet computers erroneously detected a US first strike; global catastrophe was only averted when an otherwise obscure officer named S.Y. Petrov refused to believe the data.

In contrast to German military cartography of the Second World War, which made only patchy use of aerial reconnaissance and interpreted the raw data clumsily, Soviet military cartography was impressive in every way; and it is only recently that fuller details of its scale and global reach have emerged outside of intelligence circles in the West. At its height, it is estimated that 35,000 staff were employed in cartographic work by the Soviet GUGK (Chief Administration of Geodesy and Cartography), and that more than one million separate maps were produced. These were drawn in Moscow, but printing was farmed out to special high-security factories as far afield as Kiev, Tashkent, Irkutsk, and even Khabarovsk on the Pacific coast.

The GUGK gathered information eclectically from a wide range of sources. Standard published maps by the British

ГУГК, Эдинбург = GUGK, *Edinburg (N-30-6)* (1983)

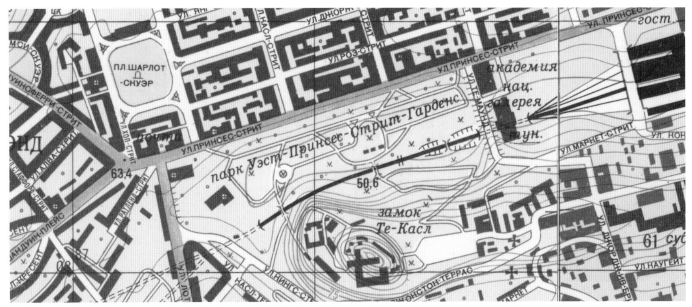

Central Edinburgh, including Charlotte Square (upper left), Princes Street gardens (centre) and Waverley station (right). The law courts and other 'administrative' buildings around Parliament Square appear in prominent purple.

Ordnance Survey and commercial publishers such as Bartholomew were often primary sources, even though in some cases these were many years old and inconsistent. But what is most striking is the large quantity of information *not* featured on these commercially available maps, which therefore must have been obtained via direct espionage or from unwitting informants on the ground. For example, contours are at 2.5 or 5-metre intervals, with spot-heights along streets shown to tenths of a metre. Most streets, significant buildings and regions are named, along with the dimensions of streets and the widths and heights of selected bridges – which would have been of particular interest in the event of an armour-led ground invasion. The map was printed in 10 colours, with buildings listed on an accompanying sheet and colour-coded – military in green, administrative in purple, industrial in black and residential in brown – making it one of the most aesthetically pleasing maps of the city ever made, despite its sinister origins and violent end-purposes. In all, the Soviets mapped more than 80 British towns, including Aberdeen, Dunfermline, Glasgow, Greenock and Kilmarnock in Scotland; Dundonians may feel insulted, yet relieved, at their exclusion.

For all Soviet military map series, the sheets follow those of the International Map of the World, a project which ironically enough was started in London in 1913. Its system allowed the globe to be completely organised, classified and brought together through 4-degree-wide latitude bands, and 6-degree-wide longitude zones. North of the equator, the bands are coded from (N)A to (N)V, while the zones are coded from 0 to 60, working eastward from 180 degrees longitude, i.e. the International Date Line. Grid square (N)N-30 therefore lies between 52 and 56 degrees north and between 0 and 6 degrees west, and is further subdivided into 144 sheets per grid square at 1:100,000.

The map also illustrates how far we use text in map-reading: although Edinburgh is instantly recognisable to those familiar with its topography, others unfamiliar with the Cyrillic alphabet will be thrown by the lack of textual clues as to what they are looking at. The mapmakers did, however, provide a phonetic transliteration of 'Edinburgh' (Эдинбург) so that Russian speakers could easily read and pronounce it.

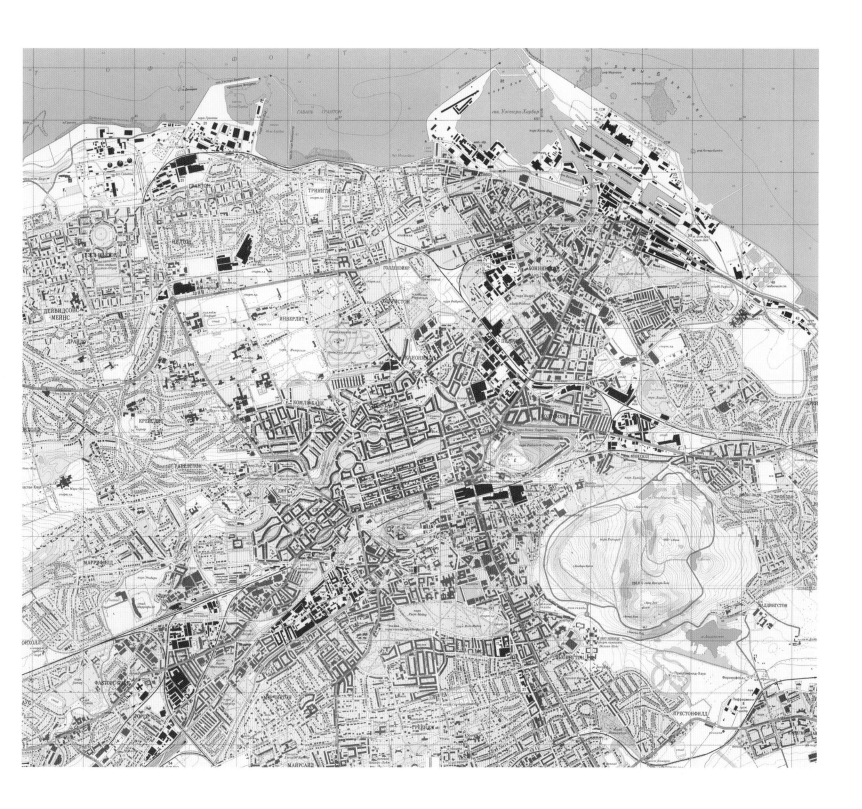

2014

Open, collaborative, volunteered digital data

Digital technologies have completely transformed cartography, not so much because of the technologies themselves, but the changes they have facilitated in source information, the people and institutions making maps, and how mapping is used and presented. Around the world in the 1980s, a number of family firms and small cartographic companies could still be found, drafting content using physical tools and printing their maps on paper, a process not fundamentally changed since the days of Blaeu in the seventeenth century [c.1610, 1647a]. Over the next three decades, however, most of these were either closed or superseded by completely different production processes, often owned by large multinational companies, with production distributed globally. Whether or not the resultant mapping is 'better' or 'worse' depends entirely on the perspective or detail chosen, and the outputs still represent or misrepresent the world as much as their predecessors, as well as continuing to reflect the concerns and priorities of the age in which they were created.

OpenStreetMap (OSM) has been described as a 'Wikipedia of Maps', allowing members of the public to collaborate to create a free, editable map of the world. The project was initiated in 2004 by Steve Coast, a computer science graduate of University College London. Most participants were initially based in the London environs, but the project quickly spread beyond the United Kingdom, and by early 2013 boasted more than a million registered contributors. It is supported by the OpenStreetMap Foundation, a non-profit organisation registered in England.

OSM mapping reflects a number of quite specific characteristics of present-day technology, cartography and society. Digital mapping technologies have evolved at a tremendous pace in the last two decades, and mapping has become ubiqui-

OpenStreetMap, Detail of central Edinburgh (2014).
© OpenStreetMap contributors

tous to a degree that was not previously possible. As feature-coded geospatial data, it can be presented in ever-changing ways and used for a wide variety of locational or navigational purposes, as well as combined or 'mashed-up' with other datasets to create new types of virtual presentations. A key prerequisite for OSM has been the widespread availability of relatively cheap GPS receivers for tracing or locating features, and the ability for contributors dispersed around the world to add or modify content instantly. At the level of motivation, OSM has been a deliberate reaction against the centralisation of ownership of digital maps and geospatial datasets and their commodification through proprietary licensing restrictions. Unlike Google or Bing map and satellite layers, OpenStreetMap data is distributed under deliberately liberal licences, with the intention of promoting its free use and redistribution. Originally this was through a Creative Commons open-content licence, though from 2012 this was replaced by an Open Database Licence from the Open Data Commons; this was done to better reflect OSM's status as a database as opposed to a work of art, a key distinction in UK law. Although the map may be the most visible result of the OSM project, the underlying data is in fact the most important long-term output.

The OSM community have also been able to synthesise information from a number of data sources into a very comprehensive topographic map. GPS receiver data still represents the largest and most reliable contribution, but the use of Yahoo satellite imagery as a backdrop since 2006, and Microsoft Bing satellite imagery since 2010, has allowed excellent overhead detail for tracing, particularly of urban areas. Out-of-copyright historical mapping has also been a useful supplementary source, whilst occasional bulk imports of topographic datasets from collaborating commercial suppliers have also helped in certain countries.

This OSM detail of central Edinburgh shows how impressive and detailed such mapping can be: significantly richer in content and features than standard commercial online sources. It clearly presents a wide range of public amenities – churches, museums, galleries, health centres, schools, postboxes and public telephones – and its sub-categorisation of retail outlets into specific types allows a more utilitarian and less overtly commercial/promotional presentation of the real world. The detailed recording of footpaths, tracks, cycleways and bus stops also stands in contrast to Google or Bing maps, which are built very much for and from the point of view of the individual motorist. In spirit, ownership, presentation and impact, therefore, OSM presents a different way of looking at the world; and it is not surprising that it is increasingly used for community-development and humanitarian purposes.

ABOVE. OpenStreetMap depiction of the junction of the Canongate with Jeffrey Street, featuring rich topographic detail and subject categorisation with symbols, and including a variety of non-commercial and commercial features.

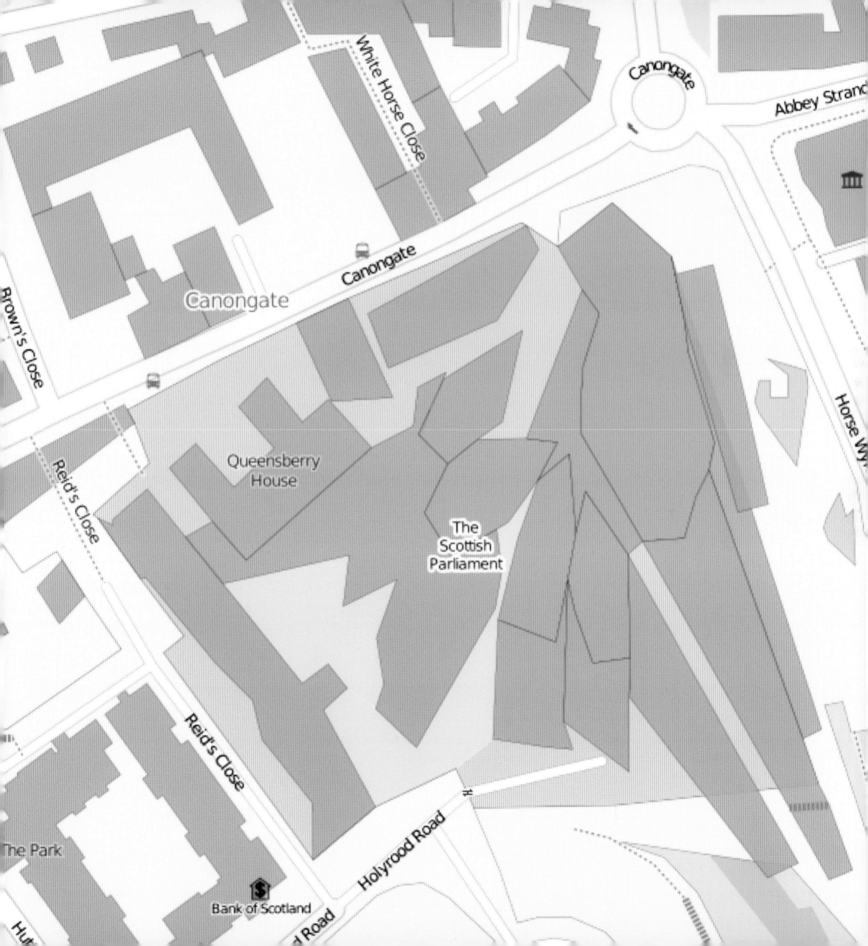

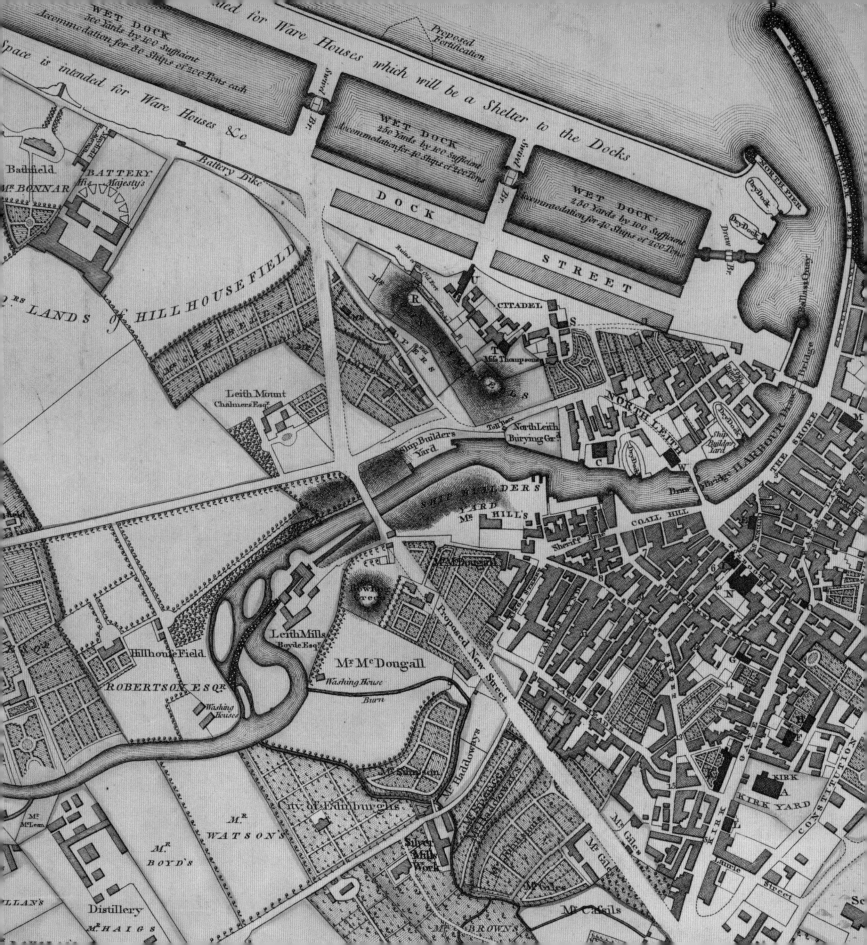

Epilogue

History without *Geography*, like a dead carkasse, hath neither life, nor motion at all.
– Peter Heylyn, *Cosmographie* (2nd edn, 1657), p. 19

Despite the manifest differences between them, the maps presented in this volume have a great deal in common, insofar as all of them are manifestations of human desires and human conflicts. To be sure, they include triumphs of insight and innovation, whether cartographic, military or commercial; perseverance; foresight; imagination; selflessness; and even simple luck. But each also tells a story of mankind's unfailing ability to demarcate differences between its members: who is a patriot and who a traitor; who is pious, who a heretic; who is an expert or an amateur, soldier or civilian, enfranchised or unenfranchised, healthy or diseased; who is equipped with the modern wonders of indoor plumbing, motor transport and electricity, and who watches them from the sidelines. To study the mapping of Edinburgh is thus to plunge oneself into the best and the worst aspects of the human character and of human life in modern times.

The next phases of both mapmaking and map use are already discernibly different, not least in that the makers and users of maps are ever more likely to be machines. Pilotless vehicles, for purposes ranging from environmental monitoring to state-sponsored assassination to pizza delivery, already exist and are increasing in numbers, geographic distribution and affordability. These machines, along with driverless cars and trains, satnavs, and other technologies yet to be invented, will increasingly communicate geographic information directly with one another. Coupled with humans' growing dependence on such technologies, and our concomitant loss of spatial awareness (even as our mental images of Edinburgh and other places become ever more closely tied to past and current cartographic representations), this 'post-human' turn may yet be remembered as the profoundest revolution in cartography's long and morally ambiguous history.

That being said, the trends of crowd-sourcing and volunteered geographic information (VGI), including innovative projects like the Mapping Edinburgh's Social History (MESH) Atlas of Edinburgh which is 'updateable on demand' by scholars, local history groups and members of the general public, point in a very different direction: one in which the ubiquity of mapping created by, for, and ultimately *about* non-specialist human beings will increasingly influence the way Edinburgh is lived in, worked in, visited and understood. As participative consumers and creators of mapping, we can therefore look forward to playing a dynamic role in the future development of Edinburgh's cartography, and of Edinburgh itself.

Further reading

Cartography

Anderson, C.J. Constructing the Military Landscape: the Board of Ordnance maps and plans of Scotland, 1689–1815. Unpublished PhD thesis, University of Edinburgh, 2010 (http://www.era.lib.ed.ac.uk/handle/1842/4598).

Barrott, H.N. (comp.), *An Atlas of Old Edinburgh*, 2nd revised edition (Edinburgh: West Port Books, 2008).

Byrom, C., 'The development of Edinburgh's second New Town', *Book of the Old Edinburgh Club*, New Series, 3 (1994), 37–61.

Cowan, W., 'Maps of Edinburgh, 1544–1851, with catalogue', *Book of the Old Edinburgh Club* 12 (1923), 209–247.

Cowan, W. and Inglis, H.R.G., 'The early views and maps of Edinburgh', *Scottish Geographical Magazine* 35 (1919), 315–330.

Cowan, W. and Boog Watson, C.B., *The Maps of Edinburgh, 1544–1929*, 2nd edition (Edinburgh: Edinburgh Public Libraries, 1932).

Cunningham, I.C. (ed.), *The Nation Survey'd: Essays on Late Sixteenth-century Scotland as Depicted by Timothy Pont* (East Linton: Tuckwell Press in association with the National Library of Scotland, 2001).

Dennison, E.P. et al., *Painting the Town: Scottish Urban History in Art* (Edinburgh: Society of Antiquaries of Scotland, 2013).

Donaldson, G., 'Map of the siege of Leith, 1560', *Book of the Old Edinburgh Club* 32 (1966), 1–7.

Elwood, S., Goodchild, M.F. and Sui, D.Z., 'Researching volunteered geographic information: Spatial data, geographic research, and new social practice', *Annals of the Association of American Geographers* 102 (2012), 571–590.

Fleet, C., Wilkes, M. and Withers, C.W.J., *Scotland: Mapping the Nation* (Edinburgh: Birlinn, in association with the National Library of Scotland, 2011).

Harris, S., 'The fortifications and siege of Leith: A further study of the map of the siege in 1560', *Proceedings of the Society of Antiquaries of Scotland* 121 (1991), 359–368.

Harris, S., 'New light on the first New Town', *Book of the Old Edinburgh Club*, New Series, 2 (1992), 1–13.

Hewitt, R., *Map of a Nation: A Biography of the Ordnance Survey* (London: Granta, 2010).

Inglis, H.R.G., 'Notes on the exhibition of the early maps of Edinburgh', *Scottish Geographical Magazine* 35 (1919), 134–136.

Laing, D.A., *Facsimile of Gordon of Rothiemay's Bird's-Eye View of Edinburgh, 1647: With an Historical Notice* (Edinburgh: W. & A.K. Johnston, 1865).

McClary, R., *The Earliest Views of Edinburgh 1544–1647* (Edinburgh: Tabula Antiqua Press, 1991).

Meade, M.K., 'Plans of the New Town of Edinburgh', *Architectural History* 14 (1971), 40–52.

Moir, D.G. (ed.), *The Early Maps of Scotland*, 2 volumes (Edinburgh: Royal Scottish Geographical Society, 1973 and 1983).

Moore, J.N., *The Historical Cartography of Scotland*, O'Dell Memorial Monograph No. 24 (Aberdeen: University of Aberdeen, Department of Geography, 1991).

Oliver, R., *Ordnance Survey Maps: A Concise Guide for Historians*, 3rd edition (London: Charles Close Society, 2013).

Robinson, A.H.W., *Marine Cartography in Britain: A History of the Sea Chart to 1855* (Leicester: Leicester University Press, 1962).

Simpson, D., *Edinburgh Displayed in a Collection of Plans and Views of the City* (Edinburgh: Lammerburn Press, 1962).

Welter, V.M., 'An undocumented plan for Edinburgh's first New Town', *Book of the Old Edinburgh Club*, New Series, 5 (2002), 107–109.

Wilson, D., 'Ancient maps and views of Edinburgh', in *Memorials of Edinburgh in the Olden Time*, Vol. 2 (Edinburgh: Hugh Paton, 1848), 201–204.

Withers, C.W.J., *Geography, Science and National Identity: Scotland since 1520* (Cambridge: Cambridge University Press, 2001).

Woodward, D. (ed.), *The History of Cartography, Vol. 3: Cartography in the European Renaissance* (Chicago and London: University of Chicago Press, 2007).

Surveyors, engravers, printers

Bendall, S. (ed.), *Dictionary of Land Surveyors and Local Map-Makers of Great Britain and Ireland, 1530–1850*, 2nd edition (London: British Library, 1997).

Bushnell, G.H., *Scottish Engravers: A Biographical Dictionary . . . to the Beginning of the Nineteenth Century* (London: Oxford University Press, 1949).

Schenck, D.H.J., *Directory of the Lithographic Printers of Scotland, 1820–1870: Their Location, Periods, and a Guide to Artistic Lithographic Printers* (Edinburgh: Oak Knoll Press, 1999).

Worms, L. and Baynton-Williams, A., *British Map Engravers: A Dictionary of Engravers, Lithographers and their Principal Employers to 1850* (London: Rare Book Society, 2011).

Edinburgh and Leith

There are many excellent general and specific histories of Edinburgh and Leith, and this selection focuses only on those that have a particular value for understanding the mapped landscape.

Bathurst, B., *The Lighthouse Stevensons* (London: HarperCollins, 1999).

Cogshill, H., *Lost Edinburgh* (Edinburgh: Birlinn, 2005).

Dennison, E.P., *Holyrood and Canongate: A Thousand Years of History* (Edinburgh: Birlinn, 2005).

Edwards, P. and Jenkins, B. (eds), *Edinburgh: The Making of a Capital City* (Edinburgh: Edinburgh University Press, 2005).

Fry, M., *Edinburgh: A History of the City* (London: Pan, 2010).

Gifford, J., McWilliam, C. and Walker, D., *Edinburgh: The Buildings of Scotland* (Harmondsworth: Penguin, 1984).

Grant, J., *Old and New Edinburgh* (London, Paris and New York: Cassell, 1880–1883).

Groome, Francis (ed.), *The Ordnance Gazetteer of Scotland: A Survey of Scottish Topography, Statistical, Biographical, and Historical* (Edinburgh: T.C. Jack, 1882–1885 and later editions).

Harris, S., *The Place Names of Edinburgh: Their Origins and History* (London: Savage, 2002).

Keir, D., *The Third Statistical Account of Scotland: The City of Edinburgh* (Glasgow: Collins, 1966).

Laxton, P. and Rodger, R., *Insanitary City: Henry Littlejohn and the Condition of Edinburgh* (Lancaster: Carnegie, 2013).

MacInnes, R., Glendinning, M. and MacKechnie, A., *Building a Nation: The Story of Scotland's Architecture* (Edinburgh: Canongate, 1999).

Mowat, S., *The Port of Leith: Its History and its People* (Leith and Edinburgh: Forth Ports in association with John Donald, 1994).

Rodger, R., *The Transformation of Edinburgh: Land, Property and Trust in the Nineteenth Century* (Cambridge: Cambridge University Press, 2001).

Smout, T.C. and Stewart, M., *The Firth of Forth: An Environmental History* (Edinburgh: Birlinn, 2012).

Youngson, A.J., *The Making of Classical Edinburgh* (Edinburgh: Edinburgh University Press, 1966)

Websites

Charting the Nation, http://www.chartingthenation.lib.ed.ac.uk/. Contains 3,500 images of early maps of Scotland, c.1560–1740.

Map History / History of Cartography Gateway, http://www.maphistory.info/. Includes a geographical listing of images of early maps.

Mapping Edinburgh's Social History (MESH), http://www.mesh.ed.ac.uk/. Constructing an online and hard-copy new Atlas of Edinburgh, AD1000–2000.

National Library of Scotland – Map Images, http://maps.nls.uk. More than 500 maps of Edinburgh, among over 80,000 maps of Scotland and beyond, c.1560–1960.

National Library of Scotland – Bartholomew Archive, http://digital.nls.uk/bartholomew/. Background information, inventories and listings of this Edinburgh firm, one of the largest commercial cartographic archives in the world.

OldMapsOnline, http://www.oldmapsonline.org/. The world's largest portal to freely available historical maps.

Our Town Stories, http://www.ourtownstories.co.uk/. Includes 18 georeferenced maps of Edinburgh from 1742 to 1915, combined with images of related collections from Edinburgh City Libraries.

Oxford Dictionary of National Biography online, http://www.oxforddnb.com/. Contains referenced mini-biographies of 59,000 British people of all eras, including scores of cartographers and publishers.

ScotlandsPlaces, http://www.scotlandsplaces.gov.uk/. A collaborative portal to maps and other geographical resources, including those in the National Records of Scotland, National Library of Scotland, and the Royal Commission on the Ancient and Historical Monuments of Scotland.

Unveiling of Britain, http://www.bl.uk/onlinegallery/onlineex/unvbrit/. Maps from the British Library, c.800–1600.

Index

Entries in **bold** indicate a map or other illustration.

Abbey Brewery 270
Abbeyhill 38, 150, 266
Abercorn, Marquis of 162
Abercrombie, Patrick 273, 274
    A civic survey & plan for the City & Royal Burgh of Edinburgh
        (1949) **272–75**
Aberdeen 21, 38, 82, 141, 261, 278, 284
Adair, John xiv, 37, 38, 43, 70
    Map of Midlothian (c.1682) **36–39**
Adam, Robert 54, 65, 89, 93, 94
Adam, William 54
Admiralty 45, 105, 106, 107, 145, 242
Advocates' Library 38, 246
aerial photography xii, **269**
Afghanistan 277
Africa 106, 193, 246
agriculture and animal husbandry 99, 158, 161, 162, 182, 185, 210, 211, 237, 238
Ainslie, John xiv, 69, 82, 97–99, 109, 110, 135, 137, 153
    Old and New Town of Edinburgh and Leith with the proposed docks (1804) **96–99**
Airport, Edinburgh International (Turnhouse) 218, 222
air defences 106, 267 *see also* Air Force, Royal
Air Force, German *see* Luftwaffe
Air Force, Royal 218, 267, 269
Albany Street 111
Albert Dock (Leith) 186, 206
alcohol 253, 254 *see also* brewers and breweries *and* distilleries *and* drunkenness
Alexander Kirkwood & Son
    see Kirkwood, Robert and James

Allane, Alexander 1–3, 6
America 70, 73, 106, 126, 141, 142, 186, 197, 245, 253, 261
American Civil War *see* Civil War, American
American Revolution *see* War of American Independence
Amsterdam 21, 25, 83
Anglicanism 29, 78, 102, 253 *see also* Episcopalianism
Anglo-Dutch War 35
Anning, Mary 218
Antiquaries, Society of 85, 133
Arbroath 117
Archer, James 94
architects 54, 89, 93, 94, 123, 129, 130, 165, 170, 181, 182, 201, 234, 262, 274
architecture 54, 61, 69, 115, 130, 173, 198, 217
Arctic, the 242
Ardrossan 99
Argyle Square 54, 66, 182, 218, 273
Argyll, Earl of 42
Argyll, Duke of 54
Army, British 49, 50, 78, 107, 157–58, 163, 266
    see also Board of Ordnance
army, English 5, 6, 9, 10, 34
army, French 9, 10, 50, 106, 162, 181, 261
army, Jacobite *see* Jacobite Rebellions
army, Scottish 6, 15, 17, 18, 34, 45, 102
Army, Territorial 186
Archers' Hall 89
Arniston Place 137
Arnott, Hugo 85
Arthur's Seat 2, 10, 34, 62, 105, 157, 159, 217, 218
artillery xii, 6, 7, 8, 9, 10, 11, 16, 17, 18, **19**, 20, 22, 45, **46**, **47**, 49, 50, 57, 58, 105, 125, 163, 186, 193–95, 267
    see also Board of Ordnance

INDEX

Assembly Rooms (George Street) 89
Assembly Rooms (Constitution Street) 127
astronomy and astronomers 62, 130, 193, 194
Atholl Crescent 229Australia 253
Ayr 47

Bacon, G.W. 171
Baillie, Alexander 69
bakers and bakeries 70, 182
Baldwin & Cradock 146
Balerno 186
Balfour, Sir James 13
Balfour, Sir Andrew 41, 42
Balmoral Hotel *see* North British Hotel
Baltic 186
Baltimore 142
Band of Hope 253
Bank Street 213
Banff and Banffshire 31, 106
banks 75, 78, 127, 153, 189
bankruptcy 121, 126, 154, 166, 198
Barbados 50
Barefoot Park *see* Bearford's Park
Barker, Robert, 88–91
    Panorama of Edinburgh from the Calton Hill (ca. 1790) 88–91
barracks *see* garrisons and barracks
Bartholomew (company) xi, xiv, 197, 205–07, 209, 213, 214, 218, 221, 225, 227, 229, 233–34, 237, 239, 242, 253, 254, 257, 258, 261, 262, 278
    Plan of Edinburgh for the Caledonian and North Western Railways' guide to the Edinburgh International Exhibition (1886) **196–99**
    Pocket Plan of Edinburgh (1889) 250
    Map illustrating Dr Harvey Littlejohn's paper 'Distribution of typhoid cases' (1891) **208–11**
    Bartholomew's Plan of the City of Edinburgh with Leith & Suburbs, reduced from the Ordnance Survey ... (1891) **204–07**, 227, 230
    Map of Edinburgh showing cases of Pulmonary Tuberculosis received at the Victoria Dispensary for Consumption and Diseases of the Chest during three years (1892) **212–15**
    The site of Edinburgh in ancient times (1893) **x, xii**
    Survey Atlas of Scotland (1895) 206, 250
    Edinburgh street tramways. Routes proposed to be cabled. (1895) **220–23**
    Plan of the Meadows and part of Bruntsfield Links showing the proposed site of the Usher Hall (1898) **228–31**
    Street Lighting. Plan of Mains (1898) **224–27**
    Orographical Plan of Edinburgh (1902) 250
    Suggested Plan for a National Institute of Geography (1902) **232–35**
    Plan of Edinburgh showing the areas proposed to be acquired and the markets and slaughterhouses proposed to be removed (1903) **236–39**
    Chronological map of Edinburgh showing expansion of the City from earliest days to the present (1919) **248–51**
    The Heart of Edinburgh (1923) **252–55**
    Scottish Zoological Park (1932) **236–39**
    Post Office Directory Map of Edinburgh and Leith (proof and finished copies) (1934) **260–63**
Bartholomew, John C. 199
Bartholomew, John George **199**, 206, 234, 249, 250, 251
Bartholomew, John 'Ian' 199
Bartholomew, John (junior) 199, 205, 206
Bartholomew, Peter 199
Bartholomew, Robert 199
baseline 62
Bath 273
Bathymetrical Survey 241–42
    Duddingston and St Margaret's Lochs (1903) **240–43**
Baxter, John 93
Baxter's Place 117
baxters *see* bakers and bakeries
Bearford's Park 77, 81
Beaufort, Francis 145
Beche, Henry De la 218
Begbie, Patrick 75
Belfast 234
Bell, Andrew 77, 78, 178
    A plan of the city of Edinburgh ... (1773) **76–79**
Bell, Benjamin 137
Bell, George 138
Bell, Joseph 137
Bell Rock Lighthouse 117
Bell, William 100–03
    Plan of the Regality of Canongate ... (1813) **100–03**
Bell's Brewery 66, 86, 270
Bell's Mills 155
Belleville 136–39
    Plan of the lands of Newington and Belleville (1826) **136–39**
Bells Wynd 31
bench-marks 173, 205
Berlin 225
Bermuda 106
Berwick-upon-Tweed 7, 10, 17, 22, 169
Berwickshire 53, 107, 218
Beugo, John 86
Bing 282

bird's-eye views 17–20, 21–24, 29–32
Bishops Land 242
Black Turnpike 198
Blacket estate 137
Blackett House (Dumfriesshire) 137
Blackfriars Monastery 2, 3, 6
Blackfriars Wynd / Street 15, 182
Blackhall 134
Blackwood, William 138, 179, 199
Blackwood's Edinburgh Magazine 138
Blaeu (publishers) xii, 26, 27, 31, 35, 37, 281
Blair, James Hunter 93
Blenheim 154
Board of Ordnance 49, 50, 53, 54, 102 *see also* artillery
bombs and bombing 265, 266, 267, 273
Bonnie Prince Charlie *see* Stuart, Charles Edward
Bonnington 10, 150
Bonnington Road 222
bookbinders 25, 246
booksellers 25, 94, 121, 154
Borthwick Wynd 31
Botanic Garden *see* Royal Botanic Garden
Bouch, Thomas 150
Bournemouth 273
Braun, Georg 21, 27
Breadalbane, Earl of 115
Breakwaters, docks and piers 47, 97, 123, 126–27, 148–50, 186, 206, 207, 266
brewers and breweries 66, 70, 86, 97, 138, 229, 270
bridewell *see* gaols
Bristo Port 33, 66, 118
Bristol 273
British Geological Survey 218
Brooke, William 135
Brougham, Henry 145
Brougham Place 198
Brougham Street 146
Brown, James 66, 86, 93
Brown, Robert 153
Brown, Thomas 93, 94, 127, 129
    Plan of the city including all the latest improvements (1793) **92–95**
    Plan of the City of Edinburgh, including all the Latest and Intended Improvements ... (1823) **128–31**
Bruntsfield 38, 229
Buccleuch, Dukes of 86, 115, 149, 150, 161
Buccleuch Place 86, 170
Buchanan, George 160–63
    Plan of the estate of Craigentinny (1847) **160–63**
burgage plots 31, 33

burgesses 31, 121
Burgh Engineer 262
Burns, Robert 86, 183
Burntisland 150, 169

cadastral maps 81–84, 97–100, 101–04
Caddell, H.M. x–xii, 218
    The site of Edinburgh in ancient times (1893) **x–xii**
Caledonian Hotel 199, 202
Caledonian Mercury 261
Caledonian Railway 142, 146, 170, 197, 199, 201, 202, 225–26, 274
    Plan of Edinburgh for the Caledonian and North Western Railways' guide (1886) **196–99**
    Proposed railway along Princes Street and station at Waverley Market (1890) **200–03**
Caledonian Station 201, 202, 274
Calton Hill 125, 129, 163, 169, 193, 217
Calton New Town 123
Cameron Toll 222
Campbell, Colen 54
Campbeltown 246
Camperdown, Battle of 137
Canaan 131
Canada 78, 106, 245
Canal Street 77, 82, 114, 150, 169, 173
cannon *see* artillery
Canongate 6, 10, 23, 25, 33, 34, 37, 38, 54, 69, 70, 101, 102, 161, 182, 183, 185, 214, 222, 225, 261, 270
    Plan of the Regality of Canongate (1813) **100–03**
Canongate Church 101
Canongate Poorhouse 214
Canonmills 70, 97, 98, 110, 142, 150, 189, 218
Canonmills Distillery 98, 110, 150
Canonmills Loch, 97, 189, 218
Cape of Good Cope 193
Caroline Park 149
Cassilis, Earl of 115
Cattle Market 237, 239
Castle Terrace 90, 129, 146, 230, 239
Causewayside 137, 138
Challenger Office (Queen Street) 241
Chambers, Robert 183
Chambers, William 78, 181, 182
Chambers Street 129, 182, 229, 230, 265
Chapman & Hall 146
Charles I, King 27, 29, 38, 177
Charles II, King 42, 45
Charles Goad (company) *see under* Goad, Charles
charts 45–48, 105–08, 149–52

INDEX

Chile 246
cholera 161, 173, 185, 210, 213
Christian Salvesen (company) 258
churches see religious buildings
City Architect 182, 262
City Chambers 38, 65, 129, 173
City Fever Hospital (High School Yards) 211
City Fever Hospital (Craiglockhart) 214
City Observatory (formerly Scientific Observatory) 130
civic improvement see improvement, civic
Civil War, American 186
Civil War, English 31, 35, 42, 47
Civil War, Marian 15, 17, 18
Civil War, Spanish 231
Claremont Street 222
Clarendon Crescent **166, 167**
Clarendon, George Villiers, Earl of 166
Clarke, W.B. 145
Clermiston 134
Clydebank 265
coachbuilders 114, 165
Coast, Steve 281
Coates Crescent 153
Coates estate 123, 153
Cockburn Association 173
Cockburn, Henry 173
Cockburn Street 173, 182
Cockerell, C.R. 129
Colinton 38
Collins, Greenvile 44–47
    Map of Leith from the North (1693) **44–47**
Commercial Street 149
Cons Close 31
Constitution Street 127
Cooper, Richard 38–39, 78, 178
Corn Market / Corn Exchange 238, 239
Corncockle Quarry, Dumfriesshire 202
Corsehill Quarry, Dumfriesshire 202
Corstorphine 26, 134, 217, 218, 257, 258, 266
Corstorphinhill 133, 257
    Plan of the Estates of Ravelston and Corstorphinhill (1826) **132–35**
Cotton, Sir Robert Bruce 5
county maps 25–28, 37–40, 69–72, 121–24
Cousin, David 180–83
    Revised plan of projected improvements of the Old Town of Edinburgh (1866) **180–83**
Cousland 217
Covenanters 34, 35
Cowgate 111, 182, 185, 246, 274

Cox, John 189
Craig, James
    Plan of the new streets and squares, intended for [the] *ancient capital ...* (1768) **72–75**, 113, 114
Craigentinny 110, 161, 162, 266
    Plan of the estate of Craigentinny (1847) **160–63**
Craigleith 214
Craiglockhart 214
Craigmillar 26, 38, 109, 266
Cramond 138
Crawford, David 133–35
Crawford, David 133, 135
    Plan of the Estates of Ravelston and Corstorphinhill (1826) **132–35**, 257
Crawford, William 135
Cumberland, William Augustus, Duke of 61
Custom House (Leith) 125

Dalgety Bay 105
Dalhousie, Earl of 115
Dalkeith 161, 217
Dalkeith Road 27, 137
Dalrymple, Alexander 106
Darnley, Henry, Lord 3, 13, 14, 15
    Bird's-eye view of the murder of Lord Darnley at Kirk o' Field church and churchyard (1567) **12–15**
Dean 129, 164–67
    Plan for Building on part of the Estate of Dean belonging to John Learmonth, Esq (1850) **164–67**
Dean Bridge 165, 166
Deanhaugh estate 123
depth-sounding 241
Deutsch, Hans 2
Dewar Place 225, 265
digital mapping 281
disease xi, 23, 41, 173, 185, 209, 210, 211, 213, 214
    see also cholera, syphilis, tuberculosis, typhoid, typhus
dispensaries 213, 214, 246
distilleries 98, 110, 150, 155, 265–66 see also whisky
docks see breakwaters, docks and piers
Dollarbeg Castle 267
Donaldson's Hospital 177
Doncaster 273
Downfield Place 266
Drumdryan estate 146
Drummond Place 98
Drumsheugh 69, 153
drunkenness 211, 254
Drury, Sir William 17–18

Dublin 9, 70, 234, 257, 273
Dublin Zoo 257
Duddingston 109, 110, 157, 159, 241, 274
    Duddingston and St Margaret's Lochs (1903) **141–44**
Duff Street 266
Dumbarton 57
Dumfriesshire 137, 202
Dunbar 34, 142
Duncan, Adam (Admiral) 137
Duncan Street 137
Dundas family of Arniston 137
Dundas, Henry (Lord Melville) 86, 115
Dundas, Sir Laurence 75
Dundas, Robert 115
Dundee 42, 141, 169, 246, 278
Dunfermline 278
Dunsapie Loch 158, 159
Dury, Theodore 50

East India Company 157
East Muir 137
Easter Road 171, 186, 222
Edgar & Co. 246
Edgar, William, 52–55, 64–67
    The plan of the city and castle of Edinburgh (1742) **52–55**
    The plan of the city and castle of Edinburgh (1765) **64–67**
EDINA Works (W. & A.K. Johnston) 171
Edinburgh & Dalkeith ('Innocent') Railway 117, 159, 221
Edinburgh & District Tramways Co. 221
Edinburgh & Dundee Railway 146
Edinburgh & Glasgow Railway 169
Edinburgh Artillery Militia 186
Edinburgh Castle 2, 5, 6, 9, 10, 17–19, 22, 23, 26, 33, 34, 37, 81–82, 89, 126, 146, 177, 193–94, 217, 233, 258
    'Lang Siege' of Edinburgh Castle (1577) **16–19**
    A Plan of Edinburgh Castle (1710) **48–51**
    Plan of Edinburgh Castle (1750) **56–59**
Edinburgh Central Library 246
Edinburgh Citizens 'No Licence' Council 252–55
    The Heart of Edinburgh (1923) **252–55**
Edinburgh Dock 206
Edinburgh 'royalty' boundary, ancient 'royalty' and extended 'royalty' 118
Edinburgh Gas Light Company 225, 270
Edinburgh Geographical Institute 138
Edinburgh International Exhibition 197
    Plan of Edinburgh for the Caledonian and North Western Railways' guide (1886) **196–99**
Edinburgh, Leith & Granton Railway 169

Edinburgh, Leith & Newhaven Railway 150
Edinburgh Medical Journal 214
Edinburgh, Perth & Dundee Railway 169, 173
Edinburgh Review 182
Edinburgh Social Union 233
Edinburgh Town Council *see* Town Council (Edinburgh)
Edinburgh, Treaty of 10
Edinburgh University Old College 13, 93, 94, 129, 182, 265
Edinburgh Zoological Park 134
Edinburghshire *see* Midlothian *and* Lothian
Edmonston & Son 246
Education Act (1872) 207
Edward, Nicol *see* Uddert, Nicol
Edward Stanford (company) 146
Edwards, Talbot 48–51
    A Plan of Edinburgh Castle (1710) **48–51**
Eglinton, Earl of 115
elevations of buildings 94, 113, 165, 182, 233
Elizabeth I, Queen 10, 13, 18
Engels, Friedrich 181
engineers xii, xiv, 5, 7, 9, 10, 17, 37, 42, 49, 50, 57, 58, 99, 102, 109, 123, 133, 142, 150, 163, 179, 242, 245, 262, 266, 277
England and English people xii, 5, 6, 9, 10, 13, 15, 17, 18, 26, 30, 34, 35, 38, 43, 45, 49, 58, 62, 69, 78, 81–83, 102, 166, 234, 265, 273
engravers 26, 38, 53, 78, 111, 122, 127, 145, 171, 199, 262
engraving xiv, 25, 26, 78, 99, 111, 113, 121, 146, 154, 178, 179, 206
Environment Society 233
Episcopalianism 2, 29, 31, 35, 78, 101, 102
Erskine, Mary 54, 134
estate maps **133–36**
Eton Terrace 166
Eyemouth 7, 10
Eyre, James 97

Fairmilehead 222
fascism 231
Ferry Road 150
feudalism 101
feuing 27, 81, 113–14, 137, 138, 146, 165, 181
Fever Hospital (Craiglockhart) 214
Fillyside Farm 162
Finland 253
fire insurance 244–47
    Fire Insurance Plan of Edinburgh (1906) **244–47**
Fire Station, Central 237
First World War 142, 163, 183, 214, 234, 251, 253, 265, 266, 269
Fish Market Wynd 31

fishermen 105, 242
fishery 163
fishmarket 31, 86, 127
Flesh Market 42, 127, 173, 237
Flodden Wall 15, 19, 33, 146
football 190
footpaths 158, 258, 282
forests see woodland
Forrest Road 146
Forth, Firth of 150, 186, 242
    Survey of the Frith of Forth (1815) 104–07
Forth and Clyde Canal 70, 109
Forth rail bridge 201, 266
fortifications 5, 7, 9, 10, 18, 22, 47, 49, 54, 57, 127, 170, 284
    see also Edinburgh Castle and Flodden Wall and King's Wall and Leith, Citadel of and ports / gates and Telfer's Wall
Fortune, Matthew 115
Fortune's Tavern see Tontine Tavern and Coffee House
Foul Burn 161, 162, 185
Fountainbridge 86, 150, 229, 237
Fourdrinier, Paul 53, 73
France and French people 9, 10, 18, 26, 38, 42, 49, 50, 53, 62, 74, 78, 106, 126, 142, 145, 153, 162,186, 193, 195, 197, 198, 225, 234, 246, 261, 266, 274
franchise see voting
Fraserburgh 265
freemasonry 86
French Ambassador's house (Edinburgh International Exhibition) 198
French Industrial Exhibition (1844) 197
Fruitmarket 169, 173 see also Vegetable Market

Galeron, Paul 232–35
    Suggested Plan for a National Institute of Geography (1902) 232–35
gaols, bridewells and prisons 75, 86, 93, 101, 177, 178
gardens 41, 61, 74, 98, 102, 109, 113, 133, 134, 146, 150, 153, 185, 205, 233, 257, 290
garrisons and barracks 6, 17, 18, 34, 47, 50, 54, 57, 102, 162, 163
gas works 142, 163, 186, 225, 270
gates see ports / gates
Gavin, Hector 78, 86
Geddes, Patrick 221, 232–35, 250, 257–58, 274
    Suggested Plan for a National Institute of Geography (1902) 232–35
Geikie, Archibald and James 217–18
German general staff (Generalstab des Heeres) 264–67
    Stadtplan von Edinburgh mit Mil.-Geo. Eintragungen (1941) 264–67
General Assembly of the Free Church of Scotland 142

General Steam Navigation Company 149
geographers xiv, 29, 37, 43, 69, 99, 206, 233, 234, 284
geology and geologists 216–19
Geological Survey of Great Britain and Ireland 218
Geological Survey of Scotland 216–19
    One-Inch to the Mile, Scotland – Sheet 32 – Edinburgh (1892) 216–19
George Heriot's (school) 33, 70, 77, 89, 94, 98, 110, 165, 177
George III, King 74, 78, 82, 99, 102
George IV, King xii, 74, 125–27, 134
George IV Bridge 118, 125, 129, 146, 198, 218
    Fire Insurance Plan of Edinburgh – Sheet 12 (1906) **244–47**
George V, King 214
George V Memorial Park 189
George Square 66, 69, 74, 86, 137, 185, 218
George Street 74, 75, 77, 90, 102, 125, 127, 254
Germany and German people xii, 1, 42, 98, 102, 163, 171, 189, 190, 206, 225, 264–67, 269, 277
Ghent 234
gibbet **26**, 38, 62
Gibbs, James 54
Gilmerton 217
Girth Cross 23
Gibraltar 127
Gladstone, Sir John, of Fasque 150
Gladstone, William 150
Glasgow 69, 78, 89, 90, 99, 117, 153, 169, 202, 230, 234, 246, 254, 266, 278
Glasgow International Exhibition (1901) 198
Glencoe 73
globes 171, 234
Goad, Charles 244–47
    Fire Insurance Plan of Edinburgh – Sheet 12 (1906) **244–47**
golf 134, 162, 207, 218, 258
Google 282
Gordon, James xi, xiii, 7, 38, 53, 237, 250
    Edinodunensis Tabulam (1647) **28–31**
    Urbis Edinae facies meridionales = The prospect of the south syde of Edinbrugh (1729) **32, 34**
    Urbis Edinae latus septentrionale = The prospect of the north syde of Edinbrugh (1729) **32, 35**
Gorgie 189, 214, 222, 238, 239, 265, 266, 274
Government, of England 13
Government, of Great Britain 27, 53, 74
    see also Parliament (Westminster)
Government, local see Town Council
Government, of Scotland 21, 138 see also Parliament (Scottish)
Government, of the United Kingdom 241–42, 254, 269
    see also Parliament (Westminster)

Gowans, Sir James 199
Grainger & Miller 150
Grangemouth 266
Granton 10, 109, 149, 150, 169, 234, 241, 262, 266
    Chart of the Firth of Forth from Queensferry to Inchkeith, showing the relative position of the proposed harbour at Granton ... (1834) **148–51**
Granton Road 150
Grassmarket 38, 69, 93, 118, 146, 217, 237, 246
Great Exhibition (London) 197
Great Fire of 15–17 November 1824 111, 246
Great Reform Act (1832) xii, 141, 142
Great Junction Street 127
Greek Revival 128–31
Greenock 246, 265, 278
Greenwich Observatory 194
Greyfriars 2, 3, 33, 54
Groome, Francis 161
Grove Square 155
GUGK 276–79
    Edinburg (N-30-6) (1983) **276–79**
gunpowder 47, 58
guns *see* artillery
Gyle 222
gymnasts and gymnastics 189, 190

Haddington, Earl of 98, 110, 142, 157, 158
Haig, James 110
Haig, John 110
Hamburg 267
Hamilton, Dukes of 115
Hamilton, Thomas 118
Hanover Street 81, 99, 113, 125
harbours 99, 117, 125, 126, 127, 149, 150, 179, 186, 241
Hardyng, John xiii
Hart, Andro 25
Haymarket 118, 169, 237, 265
Hebrew language 1, 30
Henderson, Dr John 207
Henderson Row 221
Henderson Street 207
Heriot's (school) *see* George Heriot's (school)
Hertford, Edward Seymour, Earl of 5–7
Hewitt, John (Bathymetrical Survey) 240–43
    Duddingston and St Margaret's Lochs (1897–1909) **240–43**
Hewitt, John (Time Gun) 193
High Cross *see* Market Cross
High Kirk *see* St Giles Cathedral
High Street 111, 127, 131, 171, 181, 182, 225, 237, 274

Hislop, Alexander 195
Hogenberg, Franz 21, 27
Holland 49, 126, 137
Holyrood Abbey, 7, 41, 70, 101, 102, 157, 270
Holyrood Palace 2, 3, 5, 10, 22, 23, 26, 33, 89, 102, 157, 161, 198, 270
Holyrood Park 26, 98, 110, 156–59, 185
    Plan of Holyrood Park shewing the proposed new lines of road (1843) **156–59**
Home, John 114, 135
Hondius, Jodocus, xii, 37
    A new description of the shyres Lothian and Linlitquo (1630) **24–27**
Hopetoun, Earl of 115
horses 117, 163, 169, 221, 222, 238
hospitals *see* infirmaries and hospitals
hotels 113, 199, 202
House o' Muir 237
Howell, Henry 218
Hume, David 86, 89, 113
Hume Mausoleum 89
Hutton, James 86, 218
hydrographers 45, 106, 145
Hydrographic Office 104–07
    Survey of the Frith of Forth (1815) **104–07**

Imperial Dock 207
improvement, civic 173, 181–83
    see also sanitation and sanitary reform
    Revised plan of projected improvements of the Old Town of Edinburgh (1866) **180–83**
incarceration *see* gaols, bridewells and prisons *and* workhouses
Inch & Sons 246
Inchcape Rock 117
Inchcolm 105
Inchkeith 34, 105, 149
Industrial Museum *see* National Museum of Scotland
infirmaries and hospitals 38, 54, 262
'Innocent' Railway *see* Edinburgh & Dalkeith ('Innocent') Railway
Institute of Geological Sciences 218
insurance *see* fire insurance
International Map of the World 278
Inveresk 217
Inverleith 10, 42, 70, 98, 110, 150
Inverness 21, 47, 58
Ireland and Irish people 9, 10, 17, 70, 89, 98, 106, 145, 166, 181, 218, 234, 257, 273
Irkutsk 27
irrigation meadows 158, 161, 162, 185

J Cox (company) 146
J & C Walker (company) 145
Jacobite Rebellions and Jacobites 50, 53, 57, 61, 73, 74, 82, 102, 157, 183, 250
jails *see* gaols, bridewells and prisons
Jamaica 106
James IV, King 250
James V, King 250
James VI and I, King 15, 17, 27, 29, 93
James Court 233
Janssonius, Johannes 20–23
    Edenburgum vulgo Edenburg (1650) **20–23**
Jedburgh 99, 142
Jeffrey, Lord 182
Jeffrey Street 182
Jenners 198
Jock's Lodge 162
John Bartholomew & Son
    see Bartholomew (company)
Johnson, Rowland 16–19
    Siege of Edinburgh Castle (1577) **16–19**
Johnston Terrace 118, 129, 146, 170
Johnston, W. & A.K. xiv, 82, 122, 164–71, 179–87, 192–95, 199, 262
    Plan for Building on part of the Estate of Dean (1850) **164–67**
    Johnston's plan of the City of Edinburgh (1850) 170
    Johnston's plan of Edinburgh & Leith (1851) **168–71**
    Revised plan of projected improvements of the Old Town of Edinburgh (1866) **180–83**
    Drainage (1869) **184–87**
    Plan of Edinburgh, Leith & Suburbs [illustrating the Time Gun] (1879) **192–95**
judges / judiciary *see* lawyers and judges

Kay, James
    Kay's Plan of Edinburgh ... (1836) **152–55**
Kay, John 163
Kay, Robert 93
Keith, Alexander 133, 134
Keith family 133–35
Kelso Abbey 157
Ker, Claude 211
Ker, Henry Bellenden 145
Khabarovsk 277
Kiev 277
Kilmarnock 278
Kincaid, Alexander 69, 82, 85, 86
    A plan of the city and suburbs of Edinburgh (1784) **84–87**
Kincardine 42
Kincardineshire 133

King's Bridge 129
King's Stables Road 146
King's Wall 15, 19, 146
Kinnear, Charles 201
Kirk o' Field Church 2, 3, 5, 12–15, 23
    Bird's-eye view of the murder of Lord Darnley at Kirk o' Field church and churchyard (1567) **12–15**
Kirkcaldy 198
Kirkwood, Robert and James 82, 108–15, 127, 138, 142, 154, 158, 159, 171, 205, 209, 273
    This plan of the City of Edinburgh and its environs ... (1817) **108–11**
    Plan & elevation of the New Town of Edinburgh (1819) **112–15**
Kirkcaldy, William, of Grange 17
Knox, James 122
Knox, John 10, 18, 127, 138
Koch, Robert 213

labourers 141
Laing, Alexander 93
lamps and lamp-posts *see* street lighting
Lanarkshire 61
Lancefield, Alfred 170, 183
    Johnston's plan of Edinburgh & Leith (1851) **168–71**
landowners and landownership 75, 77, 98, 110, 118, 123, 127, 133, 162, 165 *see also* cadastral maps
Lasswade 214
Lauder 142
Laurie, Gilbert 78
Laurie, John
    A plan of Edinburgh and places adjacent (1766) **68–71**
Lauriston Place 237
Laverockbank 70
Lawnmarket 18, 54, 118, 146, 233, 245
Lawson & Son 246
lawyers and judges 38, 70, 78, 86, 115, 127, 145, 173, 182, 189, 214, 246, 258, 278
Le Corbusier 274
Learmonth, Sir John 114, 154, 165, 166
Learmonth & Co. 114
Lee, Richard
    Bird's Eye View of the Town of Edinburgh (1544) **4–7**
    The plat of Lythe w' th'aproche of the Trenches therevnto (1560) **8–11**
Leeds 253
Leicester 230
Leicester Square, London 90
Leith xiv, 2, 6, 8–11, 17, 18, 26, 27, 33, 38, 42, 44–47, 53, 62, 70, 97, 98, 101, 105, 109, 117, 123, 125–27, 129, 130, 142, 149, 150,

154, 157, 161, 162, 165, 169, 170, 173, 186, 193, 206, 207, 221, 246, 261, 266, 274
   The plat of Lythe w' th'aproche of the Trenches therevnto (1560) 8–11
   Map of Leith from the North (1693) **44–47**
   Plan of the Town of Leith and its environs (1822) **124–27**
Leith & Newhaven Railway 150
Leith Branch Railway 162
Leith, Citadel of 38, 47
Leith Dock Commission 127
Leith Improvement Scheme (1880) 207
Leith Links 207
Leith School Board 207
Leith Walk 123, 125, 127, 129. 154, 222
Lennox Street 166
Lennox Street Lane 166
Leslie, John 136–39
   Plan of the lands of Newington and Belleville (1826) **136–39**
Lessels, John
   Revised plan of projected improvements of the Old Town of Edinburgh (1866) **180–83**
Lesslie, John
   A plan of the North Loch and other grounds belonging to the city of Edinburgh (1779) **80–83**
Leuchars 166
Levenhall 222
Leviathan (ship) 150
Liberton 82, 217, 221, 222
libraries 38, 82, 138, 145, 246 *see also* Advocates' Library; Edinburgh Central Library; National Library of Scotland
lighthouses 38, 46, 105, 117, 118, 127
lighting *see* street lighting
Lilliput 70
lime-kilns 105
limestone 217
Linlithgow 21
linoleum 198
lithographers and lithography 171, 185, 199, 206, 242, 249, 262
Little France 38
Littlejohn, Harvey 208–11
Littlejohn, Henry 181, 182, 186, 214, 237
   Map illustrating Dr Harvey Littlejohn's paper 'Distribution of typhoid cases' (1891) **208–11**
Liverpool 197, 273
Lizars, Daniel and William Home 78, 171, 176–79, 199, 262
   Edinburgh Geographical Atlas (c.1840) 179
   Plan of Edinburgh (1852) **176–79**
Loch Ness 242
Lochrin Distillery 110

London 5, 10, 50, 53, 62, 70, 73, 74, 75, 86, 89, 90, 98, 102, 117, 122, 126, 137, 142, 146, 149, 166, 170, 171, 173, 197, 198, 199, 206, 210, 218, 234, 242, 246, 257, 273, 274, 278, 281
London Road 117
longitude 193, 278
Longmore, John Alexander 138
Longmore Hospital for Incurables 138
Lothian 25, 27, 102, 118, 129, 146, 170, 201, 202, 230, 267, 274
   see also Midlothian
Lothian Road 118, 129, 146, 170, 201, 202, 230
Luckenbooths 29, 261
Luftwaffe 266

Macdonald Road 226
Maclennan, Duncan 254
Macrae, E.J. 163
Macraes, Affair of the 157–58
Macree, Revd Dr 138
magistrates *see* lawyers and judges
Maitland, Sir Alexander 154
Maitland, Charles 102
Maitland, William 54
Maitland Street 154, 229
Major Weir's House (Edinburgh International Exhibition) 198
Manchester 163
Manor Place 154
manufacturing 49, 54, 86, 138
Marchmont 210
markets 22, 26, 29, 30, 31, 42, 54, 85, 98, 113, 127, 146, 173, 182, 201, 237, 238, 239
   see also Corn Market, Fleshmarket, Fruitmarket, Lawnmarket, Vegetable Market
Market Cross 18, 29, 30, 261
martello tower (Leith) 105, 106, 125, 126
martello towers (general) 105–06
mathematics and mathematicians xiv, 62, 69
Mary, Queen of Scots 5, 13–15, 17–18, 38
Mary Erskine School 134
Maudsley, Sons & Field 194
Maybury 222
Meadowbank 218
Meadows, The 62, 86, 109, 146, 158, 185, 198, 199, 229, 230, 274
   Plan of Edinburgh for the ... guide to the Edinburgh International Exhibition (1886) **196–99**
   Plan of the Meadows ... showing the proposed ... Usher Hall (1898) **228–31**
Mears, F.C. xiii, 250, 251, 257
   Primitive Edinburgh (1919) **xiii**
meat trade 237, 238

INDEX

Medicine, Edinburgh School of 209
medicine 182, 209, 214 *see also* diseases
Medico-Chirurgical Society of Edinburgh 209, 214
Melville Drive 274
Melville, Lord *see* Dundas, Henry
Melville Monument 155
Melville Street 153
Merchiston Castle School 170
Mercat Cross *see* Market Cross
Methodism 78, 102, 222, 253, 254
Middle Meadow Walk 62, 146, 198, 222, 227
Midlothian 37, 69, 117, 121, 122, 149, 169 *see also* Lothian
Midlothian County Buildings 246
military maps 5–8, 9–12, 17–20, 49–52, 57–60, 265–68, 277–80
militia 98, 157–58, 163, 186
milk 161, 210, 211
mills 70, 97, 157, 185, 189
Miller, William Henry 110
ministers 25, 31, 33, 35, 37, 105, 138, 142, 145, 159, 253, 254
    see also Knox, John
Minorca 106
Minto Street 137, 138
monasteries *see* Blackfriars, Greyfriars
Montrose 163
Moray, Earl of 162, 183
Moray estate 114, 165
Moray House 270
Morningside 109, 131, 222
Morrison & Gibb 262
Morrison Street 201, 225, 274
Morton Crescent 146
Moscow 277
Mosley, Oswald 230–31
Mount Pelham 10
Mount Somerset 10, 207
Münster, Sebastian 1–3
    Alexandre Alesie Escossois D'Edinbourg (1550) xvi, **1–3**
Murchison, Roderick Impey 218
murder 3, 13, 14, 15, 17, 18, 283
Murray, Sir John 241–44
Murray, William 140–43
    Plan of Edinburgh and Leith (1832) **140–43**
Murrayfield 134, 218
Murrayfield Golf Club 134
Muses Threnodie, The 111
museums 94, 234, 282 *see also* National Museum of Scotland
Musselburgh 9, 15, 41, 142, 221, 222, 267

Napoleon I, Emperor of France 50

Napoleonic Wars 94, 97, 105–06, 121, 129, 153
National Institute of Geography 232–35
    Suggested Plan for a National Institute of Geography (1902) **232–35**
National Library of Scotland 82, 138, 246
National Monument 129
National Museum of Scotland 182, 218, 230, 265
nationalisation 202
nationalism and nationalists 102, 183, 231
navy, French 50, 106, 162, 186
Navy, Royal *see* Admiralty
Navy, Royal Scots 45, **46**
Neele, S. 122
Nelson Monument 106, 193, 194
Netherbow Port 5, 6, 33, 37, 42, 69, 89, 198
New Theatre 198
New Town 69, 70, 71, 72–79, 85, 90, 94, 97, 111, 112–15, 117, 118, 121, 127, 129, 138, 142, 163, 165, 169, 170, 185, 189, 214, 217, 221, 225, 227, 229, 234
    Plan of the New Streets and Squares, Intended for [the] Ancient Capital of North-Britain ... (1768) **72–75**
    A plan of the city of Edinburgh ... (1773) **76–79**
    Old and New Town of Edinburgh and Leith with the proposed docks (1804) **96–99**
    Plan & elevation of the New Town of Edinburgh (1819) **112–15**
    Sketch of part of the City of Edinburgh and extended Royalty (1819) **116–19**
    see also Calton New Town *and* Northern New Town
New Western Approach *see* Johnston Terrace
New Zealand 253
Newhaven 10, 70, 126, 150, 193
Newington 136–39
    Plan of the lands of Newington and Belleville (1826) **136–39**
Newington House 137
Newington Road 137
Niddry's Wynd 93
Niddry 266
Nixon, William 156–59
    Plan of Holyrood Park shewing the proposed new lines of road (1843) **156–59**
Nor' Loch 22, 23, 34, 65, 66, 77, 80–83, 185, 217, 237
    A plan of the North Loch and other grounds belonging to the city of Edinburgh (1779) **80–83**
Norway 253, 266
North Berwick 142
North Bridge 66, 69, 78, 81, 89, 90, 93, 94, 113, 114, 129, 237
North British Hotel 113, 202
North British Railway 146, 150, 162, 169, 199, 201, 202, 229
North Leith Church 125, 127

North Loch *see* Nor' Loch
North Richmond Street 86
Northern Lighthouse Board 127 *see also* lighthouses
Northern New Town 97, 114, 189
Northern Tramways Co. 221
Northumberland 9
Norway 15, 126, 253, 266
Norwich 5
Nuremberg Chronicle xiii

oceanography and oceanographers 241
Old Assembly Close 111
Old Fishmarket Close 31
Old Tolbooth Wynd 10
Oliver & Boyd 177
Oliver, Bryam 231
One O'Clock Gun 192–95
OpenStreetMap 280–83
    Central Edinburgh (2014) 280–83
Operation Revue 269
Orchardfield estate 129, 146
ordnance *see* artillery *and* Board of Ordnance *and* Ordnance Survey
Ordnance Survey xiv, 58, 61, 62, 105, 133, 158, 172–75, 188–91,
    205, 218, 242, 246, 250, 262, 265, 269, 278
    Town Plan – Five foot to the mile (1852 and 1877) 172–75
    Town Plan – Five foot to the mile (1876) 188–91
    Air Photo Mosaic (1946) 268–71
Orkney 107
Orphans' Hospital 89
Outlook Tower 234
Oxford 166
Oxgang (unit of measurement) 133

pageant 125
Paisley 246
pandas 258
panoramas 41, 53, 89–92
paper-making 198
Park Ale Stores 270
Paris 62, 142, 193, 225, 234
Paris Exposition 234
Paris, Matthew xiii
parkland 26, 86, 114, 133, 158, 159
Parliament Close / House / Square 29, 38, 78, 94, 99, 111
Parliament (Edinburgh) 102, 270
Parliament (Westminster) 7, 93, 94, 118, 127, 137, 141, 146, 149,
    198, 229, 230
Parliamentary Reform Act (1867) 141
Parthenon, the 129

Peach, Ben 218
Peebles Wynd 31
Peddie, John 20
Pelham's Battery / Mount Pelham 10, 207
penguins 258
Pennsylvania 261
Pentlands 2, 185, 217
People's Palace (Glasgow Green) 230
Percy, Henry (Earl of Northumberland) 9
Perth 38, 47, 111, 141, 169
Perthes, Justus 206
Petermann, Augustus 206
Petworth House 9
Philadelphia 70
Philadelphia Centennial Exhibition (1876) 197
Philip, Dr R.W. 213, 214
    Map of Edinburgh showing cases of Pulmonary Tuberculosis ...
    (1892) 212–15
philosophers 86, 102
photographs and photography xii, 242, 265, 269, 270, 274
physicians 42, 75, 213, 214 *see also* medicine *and* diseases
Physick Garden 41, 42, 169
piers *see* breakwaters, docks and piers
Piershill Cavalry Barracks 162, 163
Pilrig 10, 123, 222
Pilton 262
plagiarism 127
plague 1
planetarium 190, 234
planimeter 242
Playfair, William Henry 94, 123, 129, 130, 165, 170
Pleasance, the 66, 101, 102, 274
Pleasance Brewery 270
Plumstead, Derek 273, 274
    A civic survey & plan for the City & Royal Burgh of Edinburgh
    (1949) 272–75
Plymouth 273
Police Commission 170
police 78, 171, 182, 209, 225
pollutants 182, 185, 186
Polton Farm Colony 214
Pont, Timothy 25, 26, 27, 37, 284
    A new description of the shyres Lothian and Linlitquo (1630)
    24–27
Poor Law Commissioners 173
poorhouses 38, 41, 54, 102, 127, 198, 214
population growth 161, 181–82, 274
Port Hopetoun 110, 127
Portobello 131, 142, 162, 163, 217, 266, 274

ports / gates 22, 23, 33, 106, 225, 265
Portsburgh 53, 118
Portsmouth 106
Portugal 153, 251
Post Office 115, 194 *see also* Post Office Directories
Post Office Directories 179, 185, 190, 195, 207, 214, 239, 249, 253, 260–63
    Post Office Directory Map of Edinburgh and Leith (proof and finished copies) (1934) **260–63**
Poultry Market 173
poverty 99, 186, 210, 211
Powburn 38
Poyais 153
Prague 198
Princes Street 74, 75, 77, 90, 112–15, 199, 200–03, 274–75
Princes Street Gardens 113, 200–03, 274
Pringle, J.W. 140–43
    Plan of Edinburgh and Leith (1832) **140–43**
printing xiii, xiv, 25, 86, 111, 138, 145, 171, 198, 199, 206, 242, 261, 262, 277, 281 *see also* publishers *and* engravers *and* engraving *and* lithographers and lithography
prisons *see* gaols, bridewells and prisons
professions 78
professors 1, 94, 171, 186, 193, 209, 210
propaganda 231, 253
Ptolemy, Claudius xiii
Public Works Office 262
publishers 18, 21, 25, 78, 121, 138 146, 154 171, 179, 182, 195, 199, 205, 261, 262, 278
Pullar, Fred 242
Pullar, Laurence 242

quarrying 58, 158, 199, 241, 257
Quarryholes 38
Quebec 106
Queen Street 74, 77, 97, 241
Queensferry Road 69, 134, 153, 166
Queensferry 149, 165

radar 266
Raeburn estate 165
railways 89, 109, 117, 146, 150, 163, 169, 170, 173, 177, 197, 198, 201, 237, 239, 245, 265, 266, 274
    see also Caledonian; Edinburgh & Dalkeith ('Innocent'); Edinburgh & Dundee; Edinburgh & Glasgow; Edinburgh, Leith & Granton; Edinburgh, Leith & Newhaven; Edinburgh, Perth & Dundee; Leith & Newhaven; Leith Branch; North British
railway locomotives 169, 197

Ramsay, Allan 82
Ramsay Gardens 233
Ramsey, Andrew 218
Ramsden, Jesse 62
ravelin 50
Ravelston 133, 134, 135, 257
    Plan of the Estates of Ravelston and Corstorphinhill (1826) **132–35**
Ravelston Dykes Road 135
Ravelston estate 132–35
Ravelston Golf Club 134
Ravelston House 133–34
Ravelston Tower 133
Reagan, Ronald 277
rebellions 10, 50, 53, 57, 61, 73
Reclus, Élisée 234
recreation 189, 230
Redhall 198
Regent Bridge, 78, 117, 118
Regent Road 117
Regent Terrace 165
Regent's Park Zoo, 257
Register House 78, 90, 114, 142, **179**
religion 29, 31, 43, 78, 86, 142, 230, 253, 254
religious buildings 2, 3, 5, 7, 10, 13, 15, 23, 30, 35, 41, 47, 57, 75, 78, 81, 89, 90, 101, 102, 106, 123, 125, 127, 130, 170, 178, 194, 201, 205, 230, 238, 239, 246, 282
religious extremism 17, 23, 31
Renaissance xiii, 3, 7, 234, 249, 250
Rennie, John 99, 109, 126
Representation of the People (Scotland) Act 1832 *see* Great Reform Act (1832)
Representation of the People Act (1884) 141
reservoirs 82, 94, 102, 127
Restalrig 10, 161
Revolution, American *see* War of American Independence
Revolution, Covenanting 45 *see also* Civil War, English
Revolution, French 106 *see also* Napoleonic Wars
Revolution, Glorious *see* Revolution, Williamite
Revolution, Industrial 110
Revolution, Williamite 43, 73, 75
Revolutions of 1848 141
Richmond Street 86
rifle ranges 159
Rig Street 146
Rigg, James Home 146
Ritchie, F.J. & Sons (clockmakers) 194
roads and routeways 17, 97, 165, 274
road-traffic 274

Rocheid, James 98, 110
roperies 127
Roseburn 186
Ross, John Cockburn 153
Roxburgh 102
Roy, William xiv, 57, 58, 61, 62
    Military Survey of Scotland (1752–55) 60–63
Royal Academy 179
Royal Bank Close 245
Royal Botanic Garden 98, 150
Royal Circus 97
Royal High School 129–30, 171, **179**
Royal Infirmary 54, 262
Royal Mile *see* High Street *and* Canongate
Royal Navy *see* Admiralty
Royal Observatory (Blackford Hill) 194
Royal Patent Gymnasium 189, 190
Royal Riding Academy 201
Royal Scottish Academy 130, **179**, 201
Royal Scottish Geographical Society 249
Royal Society of Edinburgh 242
Royal Society of London 195, 242
royalty 27, 73, 102, 117, 118 *see also* specific monarchs
Royston 149
Russia 45, 126, 277–78 *see also* GUGK
Rutland Square and Rutland Street 154, 201, 202, 274

St Andrew Square 33, 75, 77, 78, 171
St Andrew Street 77, 99, 111, 169
St Ann's Brewery 270
St Anne's Street 114
St Anthony's Chapel 62, 157
St Bernard's Football Club 190
St Cuthbert's Church 2–3, 78, 81, 89, 131
St George's Free Church 201
St Giles Cathedral 2, 3, 6, 10, 23, 26, 29, 30, 33, 37, 74, 86, 89, 118, 173, 177, 194
St James Square 115, 178
St Leonards 159, 162, 169
St Margaret's Chapel **34**
St Margaret's Loch 158
    Duddingston and St Margaret's Lochs (1903) **141–44**
St Mary's Street 182
St Mary's Wynd 182
St Monans Wynd 31
St Paul's Cathedral 229
St Petersburg 142, 225
Salisbury Crags 34, 62, 125, 158, 217, 218
Salisbury Centre 138

Salisbury Road and Salisbury Place 138
Salvation Army 253
Salvesen, Lord 258
sanctuary 43, 102, 157
Sandby, Paul 58, 62
sanitation and sanitary reform 173, 181, 182, 183, 190, 209, 210
satellite 53, 282
satnavs 283
Saughton 38
sawmill 265
Saxe-Coburg Place 129
Schedel, Hartman xiii
schools 37, 38, 54, 66, 78, 85, 102, 121, 127, 129, 134, 170, 171, 177, 207, 209, 211, 230, 234, 282
science 197, 218, 234, 242, 281
Scotland Street Railway Station 189
Scotland Street Tunnel 150, 163, 169
Scotsman, The (newspaper) 159, 190, 227, 231
Scott, Sir Walter 78, 125, 163
Scott Monument 177
Scottish Geographical Magazine xi, xiii, 234, 249, 250
Scottish Life Assurance Office 221
Scottish Lighthouse Service 118
    see also lighthouses
Scottish Marine Station 241
Scottish National Gallery 130
Scottish Naval and Military Academy 201
SDUK *see* Society for the Diffusion of Useful Knowledge
Seafield 110
Second World War 106, 134, 150, 159, 197, 213, 214, 222, 254, 265–67, 269, 270, 273, 277
servants 13, 15, 138
Seven Years War 78, 261
sewers 162, 185, 186
    Drainage (1869) **184–87**
Seymour, Edward (Earl of Hertford) *see* Hertford, Edward Seymour, Earl of
Shakespeare Square 78, 90, 115
Shakespeare, William 17
Shandwick Place 153, 229
Sheffield 209, 273
Sheriff Court 218, 246
Shetland 107, 198, 199
shipwrecks 38
shipyards 127
shops 29, 94, 114, 246
Shore Street 125
Shrubhill 222
Sibbald, Robert 29, 37, 38, 41, 42, 43

Signal Tower 125
Signet Library 246
Skinner, William 50, 57, 58
    Plan of Edinburgh Castle (1750) **56–59**
slaughterhouses 229, 230, 237, 238, 239 *see also* meat trades
    Plan of Edinburgh showing the areas proposed to be acquired and the markets and slaughterhouses proposed to be removed (1903) **236–39**
slavery 157
Slezer, John 37, 38, 41, 42, 43
    The North Prospect of the City of Edenburgh (ca. 1710) **40–43**
slums 101, 130, 181, 182 *see also* improvement, civic *and* sanitation
smallpox 13
Smith, Adam 86, 102
Smith, Charles Piazzi 193–95
Smith, James 54
Smith, Thomas 117
Snow, John 173, 210
Society for the Diffusion of Useful Knowledge *Edinburgh* (1834) **144–47**
Somerset's Battery *see* Mount Somerset
South Africa 106, 246
South Bridge 66, 75, 78, 86, 93–94, 182
South Leith Church 127
South St Andrew Street 111
Spain 6, 50, 126, 231
Spanish Succession, War of 50
Spence, Sir Basil 274
station 77, 150, 159, 162, 169, 170, 173, 189, 199, 201, 202, 225, 226, 234, 237, 241, 265, 274
steamships 149, 150, 177
Steel, James 146, 166
steeples 30, 33, 89, 90, 194
Stenhouse 222
Steven Laws Close 31
Stevenson, Robert 109, 117, 118, 149, 150, 163, 179, 273
    Sketch of part of the City of Edinburgh and extended Royalty (1819) **116–19**
    Chart of the Firth of Forth ... showing the ... harbour at Granton ... (1834) **148–51**
Stevenson, Robert Louis 117, 163
Stirling 7, 57, 242
Stirlingshire 53
Stockbridge 97, 123, 165
Stockdale, Harrison & Sons 230
Stonehaven 266
street lighting 90, 117, 142, 173, 190, 198, 205, 225, 226, 227
    Street Lighting. Plan of Mains (1898) **224–27**
Stuart & Co. (Edinburgh International Exhibition) 198

suffrage *see* voting
sugar-refining 198
Sunbury Distillery 155
surgeons 86, 137, 138, 182, 209
Surgeons' Hall 66
Surrey 245
surveying and surveyors xi, xiv, xv, 5, 17, 25, 37, 38, 43, 45–46, 50, 53, 57, 58, 61–62, 66, 69, 70, 81, 82, 98, 99, 107, 109, 113, 122, 123, 127, 129, 133, 135, 138, 142, 145, 154, 158, 163, 170, 172–73, 217–18, 233, 241–42, 245, 246, 250, 269–70, 274
Strozzi, Piero di 9
Stuart, Charles Edward (Bonnie Prince Charlie) 157
Sweden 2, 102, 126
Switzerland 273
Synod Hall 90, 230, 239
syphilis 13

tailors 54, 114
Tait, J. 164–67
    Plan for Building on part of the Estate of Dean (1850) **164–67**
Tanfield 142, 262
tanners 82
Tashkent 277
Tay Bridge Disaster of 1879 150
Telfer's Wall 146
Telford, Thomas 109–10, 165
temperance movement 253–54
temples 130, 198, 177, 234
tenements 181, 211
Teviot Row 86
Thatcher, Margaret 277
Theatre Royal 115
theodolites 61, 62
Thomson, Charles (engraver) 127, 129
    Plan of the Town of Leith and its Evirons with its Intended Improvements (1822) **124–27**
Thomson, Charles Wyville (academic) 241
Thomas, George 104–07
    Survey of the Frith of Forth (1815) **104–07**
Thomson, James 74–75
Thomson, John xiv, 120–23, 134, 179
    A New General Atlas (1817) 121, 134
    Cabinet Atlas (1819) 121
    Edinburgh School Atlas (1820) 121
    Northern Part of Edinburghshire (1822) **120–23**
    Atlas of Scotland (1832) 121–23, 134
tightrope-walking 30
Tillie & Turner 246
time-ball *see* time-gun

time-gun 193, 194
Tolbooth (Canongate) 33, 101
Tolbooth (Leith) 127
Tolbooth (St Giles) 6, 29, 86, 198
Tollcross 109, 118, 222, 237
Tontine Tavern and Coffee House 115
tourism and tourists 65, 150, 177, 178, 201, 237, 267, 258
Tower Street 125
tower-houses 25, 26, 37, 38, 62, 102, 133, 249
Town Council (Leith) 97, 127, 186
Town Council (Edinburgh) 27, 31, 69, 70, 77, 81, 82, 98, 102, 109, 113, 158, 161, 170, 182, 202, 214, 225, 227, 229–31, 237–39, 246, 262, 274
    Plan of the Meadows and part of Bruntsfield Links showing the proposed site of the Usher Hall (1898) **228–31**
    Street Lighting. Plan of Mains (1898) **224–27**
    Plan of Edinburgh showing the areas proposed to be acquired and the markets and slaughterhouses proposed to be removed (1903) **236–39**
town planning 233, 234, 250, 269
trace italienne 7, 9, 10
Trades Maiden Hospital 54, 134
tradesmen 141
trains 150, 169, 170, 173, 199, 214, 283
trams 197, 221, 222
    Edinburgh street tramways. Routes proposed to be cabled. (1895) **220–23**
traverses 18, 61
trees 13, 133, 205 *see also* woodland
triangulation 37, 61, 62
Trinity (community) 149, 150, 169
Trinity Chain Pier 150
Trinity College Church 2, 3, 33, 41, 170
Trinity Crescent 150
Trinity Hospital 238
Trinity House 46
Trinity House (Leith) 127
Tripp, Sir Herbert Alker 274
Tron Kirk 30, 33, 37, 78, 93, 111
tuberculosis 213, 214, 251
    Map of Edinburgh showing cases of Pulmonary Tuberculosis ... (1892) **212–15**
turntables **168**, 169, 170
typhoid 209, 210, 211, 213
    Map illustrating Dr Harvey Littlejohn's paper 'Distribution of typhoid cases' **208–11**
typhus 161, 173

Ubaldini, Migliorino 9, 18

Uddert, Nicol 93
unicorns 198
Union Canal 109, 110, 111, 127, 229
unionists 74, 98
Upper Gray Street 138
Usher Hall 228–31
    Plan of the Meadows ... showing the proposed site of the Usher Hall (1898) **228–31**

Valleyfield Street 237
velocipede 189, 190
Vegetable Market 201
Venezuela 246
vennels *see* wynds
Versailles 229
Victoria Dispensary for Consumption 213
Victoria Dock 186, 206
Victoria Hospital for Consumption (Craigleith) 213
Victoria, Queen 161, 166, 199
Victoria Street 118, 129, 146, 218, 246
Vienna 213
views 33–36, 41–44, 89–92, 123, 124
villages 7, 25, 26, 38, 62, 138, 165
villas 75, 78, 134, 138, 185
Villiers, George (4th Earl of Clarendon)
    see Clarendon, George Villiers, Earl of
violins 198
volcano 217
voting 111, 141, 142, 230, 253, 254

W. & A.K. Johnston *see* Johnston, W. & A.K.
Walker, William 153
warehouses 45, 246
Warrender Park Road 222
warships *see* Navy, Royal *and* Navy, Royal Scots *and* navy, French
War of American Independence 106, 157
War of 1812 106
Washington, D.C. 225
Water of Leith 10, 17, 26, 27, 38, 62, 109, 126, 161, 162, 165, 185–86
Waterloo Place 117, 118
Watson, David 58, 61
Watson, E.R. 240–43
    Duddingston and St Margaret's Lochs (1897–1909) **240–43**
Watson, James
    ... Plan of the city including all the latest improvements (1793) **92–95**
Waverley Bridge 169, 173
Waverley route 170

Waverley Station 77, 150, 169, 173, 199, 201, 202, 274
weighbeam 30
wells 18, 89
West Bow 118, 146, 245, 246
West End 114, 118, 123, 161, 166, 227, 265
West Port 118
Western Approach 146 *see also* Johnston Terrace
whaling 258
wharfs 207
whisky 58, 198, 266 *see also* distilleries
Wikipedia 281
Williamson, Peter 261, **262**
Wilson, James 218
Windsor 90
Winnelstraelee 38
Wittenberg 1
Wood, John xiv, 94, 129, 130, 131, 153, 262
    Plan of the City of Edinburgh, including all the latest and intended improvements... (1823) **128**–31
    Town Atlas of Scotland (1828) 129

woodland 25, 26, 62, 134, 158, 257, 258
woods *see* trees *and* woodland
Woods and Forests, Board of 158
Woolwich 58
workhouses 102
working class 163, 214
World War I *see* First World War
World War II *see* Second World War
Wren, Christopher 73, 75
wynds 6, 15, 31, 33, 53, 54, 93, 102, 111, 131, 173, 182, 198, 246

yacht 45, 125
Yahoo 282
York Place 222
Youngson, A.J. 70, 85

Zeppelin 266
zodiac 197
zoos 134, 257, 258